SWANN'S WAY

SWANN'S WAY

THE SCHOOL BUSING CASE AND THE SUPREME COURT

Bernard Schwartz

New York   Oxford
OXFORD UNIVERSITY PRESS
1986

Oxford University Press

Oxford   New York   Toronto
Delhi   Bombay   Calcutta   Madras   Karachi
Petaling Jaya   Singapore   Hong Kong   Tokyo
Nairobi   Dar es Salaam   Cape Town
Melborne   Auckland

and associated companies in
Beirut   Berlin   Ibadan   Nicosia

Published by Oxford University Press, Inc.,
200 Madison Avenue, New York, New York 10016

Oxford is the registered trademark of Oxford University Press

Library of Congress Cataloging-in-Publication Data
Schwartz, Bernard, 1923–
Swann's way.
Bibliography: p.
Includes index.
1. Swann, James E.—Trials, litigation, etc.
2. Charlotte-Mecklenburg Board of Education—Trials, litigation, etc.
3. Busing for school integration—Law and legislation—United States.
I. Title.
KF228.S9S39   1986   344.73'0798   85-29849
ISBN 0-19-503888-6   347.304798

Printing (last digit): 9 8 7 6 5 4 3 2 1

Printed in the United States of America
on acid-free paper

AS EVER,
FOR AILEEN

Sources and Acknowledgments

This book is based upon both oral and documentary sources. The oral sources were personal interviews. I interviewed members of the Supreme Court and former law clerks, as well as Judge James B. McMillan, Julius L. Chambers, Benjamin S. Horack, and William J. Waggoner. Every statement not otherwise identified was made to me personally. I have tried to identify the statements made to me by different people, except where they were made upon a confidential basis. In the latter case, I have given the position of the person involved, but not his name. This book could never have been written without the cooperation of those who shared their time and experience so generously with me.

The documentary sources are conference notes, docket books, correspondence, notes, memoranda, and draft opinions of Justices. The documents used and their location are identified, except where they were made available upon a confidential basis. In the latter case, I have tried to identify the documents, usually by title and date, in the text. Most of these documents have never before been published.

I have been afforded generous access to the papers of the Justices and gratefully acknowledge the help given by the Manuscript Division, Library of Congress; by the Mudd Manuscript Library, Princeton University; and by the Harvard Law Library, as well as the permission given by the latter two collections to quote from the papers of Justices Harlan and Frankfurter.

I also wish to acknowledge the efforts of my editor, Susan Rabiner, the staff of Oxford University Press, and my literary agent, Gerard McCauley, the generous support of Dean Norman Redlich and the New York University School of Law, the work of my tireless secretary, Mrs. Barbara Ortiz, and help for incidental research expenses by the Charles Ulrich and Josephine Bay Foundation (through Mrs. Raymonde Paul), as well as the example of a hardworking author furnished by my son, Brian.

New York City   B.S.
September 1985

Contents

SWANN'S WAY

1

The Education of a
Southern Judge

"I am not infallible," says Sam J. Ervin, Jr., "and . . . conclusive evidence of that fact is to be found in my recommendation of Jim McMillan for the District Judgeship."[1]

The former senator was referring to James B. McMillan, the federal district judge who, in *Swann* v. *Charlotte-Mecklenburg Board of Education*,[2] first ordered extensive busing as part of a school desegregation plan. McMillan himself concedes that "Senator Ervin . . . in effect appointed me to my present job."[3] But when the senator secured the appointment in 1968, neither he nor the nominee's many other supporters had reason to expect McMillan to be anything other than the typical southern judge. After all, he had grown up as one of them and should have been the last person to disrupt the established order in what was still the most conservative part of the country. "Nobody," says the *Charlotte Observer*, "expected anything too radical from 51-year-old James McMillan, a widely respected Charlotte lawyer and Presbyterian elder, who belonged to all the right clubs and was a ranking member of the downtown establishment."[4]

A Model Southern Attorney

Born in 1916 and raised on the family farm in a rural county, Jim McMillan came from a long line of native North Carolinians and still proudly notes, "I had two grandfathers who . . . fought on the 'right' side in the Civil War against the Yankee oppressors."[5]

McMillan's background as what he terms "a traditional Southerner with a pride in the South" was scarcely affected by his four years at the

3

University of North Carolina and three at Harvard Law School, neither of which, he later noted, did much "to disturb my acceptance of apartheid as a way of life."[6] After service in the navy during World War II, McMillan went into practice and became one of his state's most respected lawyers—serving as president of the State Bar Association in 1960–61 and as a member of the North Carolina Courts Reform Commission in 1963.

In 1961, while he was president of the state bar, McMillan went to Chapel Hill and delivered a talk on race relations to the North Carolina Law Review Association. "May we forever be saved," the future judge declared, "from the folly of . . . the wild extreme of requiring that students be transported far away from their natural habitat so that some artificial 'average' of racial balance might be maintained."[7]

That statement, according to McMillan, "expressed my feelings" at that time.[8] "Five years later, my factual education began," he declares, "and my uninformed 1963 Olympian certainty about 'bussing' had to give way under the hard light of fact."[9]

McMillan himself asserts that his transformation from traditional southern lawyer to one of the unsung heroes of the civil rights movement was less a matter of metamorphosis than education. "Either you go by your ignorance," he told me, "or you go by what you learn. I elected to learn." Before the trial in the *Swann* case, "I didn't know what was going on. I had the comfortable notion that this was as good a school system as there was in this part of the world." But as he heard the evidence in the case developed, "reluctantly, but with increasing understanding, I listened and learned, and recorded what I learned."[10]

By now the notion that Justice Holmes satirized, of law as "a brooding omnipresence in the sky,"[11] has been completely discredited. But most people still do not realize how creative the role of the judge can be in the making of law. In the federal district court particularly, the individual judge to whom a case is assigned can play a decisive role in its outcome and importance. Educated in the realities of the Charlotte, North Carolina, school system, Judge McMillan would play just such a role in the school busing order that was issued in the *Swann* case.

The judge himself is short and slight, with a dour countenance that bespeaks his Scottish ancestry. His voice had been slightly altered after he broke his nose in a Golden Gloves tournament in Charlotte before the war. McMillan speaks with a decisiveness that belies the normal softness of his speech. "There was no foolishness about him even as a teenager," said a friend who had ridden with McMillan 26 miles a day on a bus to high school.

On the Firing Line

Not too long ago, federal judges in the South, like Judge McMillan, led an unruffled existence. Their decisions, except for the occasional criminal case, were scarcely noticed outside legal circles. The federal judge was a pillar of the community. He could expect to serve out his career to the accompaniment of universal esteem; at the occasional banquet in his honor, he would be the object of widespread tributes. And when he retired, he could look forward to encomia in the press extolling his dedicated public service.[12]

All that changed after the Supreme Court decisions in *Brown v. Board of Education*.[13] The federal courts now became the storm center of the emotionally charged conflict that tore southern society apart.

In its first *Brown* decision in 1954,[14] the Court held that school segregation was unconstitutional; but it did not make any provision for enforcement of the constitutional principle. That was left for the second *Brown* decision a year later.[15] In *Brown II*, the Court turned over the question of enforcement to the lower courts. Chief Justice Warren's *Brown II* opinion affirmed the discretion of district court judges to frame enforcement decrees in individual desegregation cases. The Warren opinion gave the district judges only broad guides—the most important of which was that they were to "enter such orders and decrees . . . as are necessary and proper to admit [blacks] to public schools on a racially nondiscriminatory basis *with all deliberate speed.*"[16]

The phrase "all deliberate speed" meant that there would be no immediate enforcement of the *Brown I* principle. Instead, the oxymoronic formula indicated that the district courts would provide for gradual desegregation. At a minimum this meant delay in the vindication of the right to attend desegregated schools. The result was a gap in the fulfillment of the *Brown* mandate, one that might have been avoided had the Supreme Court declared immediate enforcement in *Brown II*. As will be seen in Chapter 3, Chief Justice Warren and his colleagues never intended "all deliberate speed" to countenance indefinite delay. Yet that is just what it did do in all too much of the South.

But *Brown II* meant more than delay in desegregation enforcement. It also meant that the southern federal judges were to be the fulcrum upon which *Brown* enforcement turned. Judge McMillan and his fellow district judges were placed right on the firing line in the post-*Brown* struggle to secure a unitary school system. Whatever their personal views on segregation,[17] the federal judges were bound by the Supreme Court decisions and acted, albeit gradually, to order desegregation in southern school districts. This in turn made many of them outcasts in their own communities. After Judge McMillan's school busing order, we shall see, he became the target of abuse and a virtual pariah in Charlotte. No wonder the leading study of the role of southern federal judges in school desegregation cases was entitled *Fifty-Eight Lonely Men*.[18]

The federal judges were not the only ones placed on the firing line in the *Brown* enforcement struggle. Even more exposed were the attorneys who brought desegregation suits in the post-*Brown* years. All too often they fought for their clients at great personal risk. Most of them were brought into their cases by the National Association for the Advancement of Colored People (NAACP). Ever since its founding in 1909 the NAACP had served as a "legal lifeline"[19] for southern blacks. Particularly after the setting up of the NAACP Legal Defense Fund under Thurgood Marshall in 1939, most of the litigation to vindicate civil rights was brought with NAACP cooperation.

According to one of the *Swann* school board attorneys, "the thing that you've got to keep in mind is that this case was handled out of Columbus Circle,"[20] then the Legal Defense Fund headquarters. This was an overstatement. The NAACP has certainly been the catalyst for civil rights suits. It has provided financial help, moral assistance, and legal counsel in most desegregation cases, from *Brown* itself to the present day. But it is an exaggeration to say that all there cases were "handled out of" the Legal Defense Fund offices. The primary role in them was played by the local lawyers, and they suffered abuse and even physical violence for their efforts. "As a result of their court actions on behalf of NAACP cases, the lives of some have been endangered."[21]

To much of the white South during the post-*Brown* years the NAACP became a favorite bête noire. As the poet Langston Hughes put it, "the southern states [have] been busy since 1954 . . . seeking 'legal' ways of putting the NAACP completely out of business."[22] Laws were passed throughout the South to harass and intimidate the organization and its members. It was not until the most extreme of them were invalidated by the Supreme Court that the NAACP was able to operate freely in the desegregation effort.

Even in the Warren Court, the key decisions on the southern attempts to put the NAACP out of business almost went against the organization. Only chance, in the form of illness that forced the retirement of two Justices, led to the final decisions in favor of the NAACP.

In the 1963 case of *NAACP v. Button*,[23] the NAACP had challenged a Virginia statute that amended the laws on improper solicitation of business by attorneys to include agents for an organization which retained a lawyer in connection with an action to which it was not a party and in which it had no pecuniary interest. The Virginia Supreme Court had included the NAACP's activities within the statutory ban on improper solicitation of legal business. If upheld, this would have ended the NAACP's legal role in desegregation cases.

At the Supreme Court conference in November 1961, both the Chief Justice and Justice Black led the argument for reversal. The latter stressed that "this is but one of a number of laws passed as a package designed to thwart our *Brown* decision—a scheme to defeat the Court's order." However, despite the fact that the law's purpose was, in the Warren phrase, "to circumvent *Brown* obviously," the conference voted five to four to uphold it. The majority accepted the argument made by Justice Frankfurter. "I can't imagine a worse disservice," he declared at the conference, "than to continue being the guardians of Negroes." He argued that the statute was a valid exercise of the state's regulatory power over the legal profession. "There's no evidence here," Frankfurter asserted, "that this statute is aimed at Negroes as such."

A few weeks later, Justice Frankfurter circulated an opinion of the Court affirming the Virginia court. Justice Black wrote a dissent stressing that the statute's legislative history showed that it had been enacted with the racially discriminatory purpose of precluding effective litigation on behalf of civil rights. Though the case had thus been "decided" during the 1961 Term, before the decision could be announced, Justice Whittaker retired because of ill health and Justice Frankfurter became incapacitated after a stroke. The case was then set down for reargument in the next term.

At the October 9, 1962, conference after the reargument, Justices White and Goldberg, who had replaced Justices Whittaker and Frankfurter, voted to strike down the law. This changed the final decision to one in favor of the NAACP.

There was a similar scenario in *Gibson* v. *Florida Legislative Investigation Committee*,[24] also decided in 1963. The committee was set up by the Florida legislature to investigate the Miami NAACP branch. Gibson, president of the branch, was ordered to appear and bring with him records of members and contributors. At the committee hearing, the chairman said that the inquiry was into Communist infiltration into the NAACP. Gibson refused to produce the records. He was adjudged in contempt and sentenced to six months' imprisonment and a $1,200 fine.

At the December 1961 conference, the vote was once more five to four to affirm the conviction, with Justices Frankfurter, Clark, Harlan, Whittaker, and Stewart for affirmance. But again Whittaker's retirement and Frankfurter's collapse caused the case to be set for reargument in the 1962 Term. At the October 12, 1962, conference after the reargument, the Chief Justice and the others who had participated in the earlier conference voted as they had the first time. Justice White, who had taken the Whittaker seat, voted for affirmance. However, Justice Goldberg, the Frankfurter successor, voted the other way. That made the final decision a bare majority for reversal.

The Supreme Court thus ultimately rebuffed the southern efforts to bar the NAACP from its crucial role in desegregation cases. The *Button* and *Gibson* decisions allowed the Legal Defense Fund to continue its support in cases like *Swann*. But it was a close-run thing. Had Justices Frankfurter and Whittaker been able to cast the deciding votes, both decisions would have gone the other way.

Charlotte-Mecklenburg

Charlotte is a bustling city in the central Piedmont region of North Carolina, between the Appalachian mountains and the Atlantic shore. Located midway between Atlanta and Richmond, it is also the largest community between those two cities. Surrounded by a pleasant rolling countryside, it is both a trading and manufacturing center—the commercial hub of a large and relatively prosperous area. It has a long and proud history. When Lord Cornwallis briefly occupied the town in 1780, the hostile reception he met led him to call it "the hornet's nest," ever since the city's official emblem. At the time of the *Swann* case, its population was about a quarter of a million; with the surrounding Mecklenburg County, its metropolitan area had some 100,000 more.

North Carolina has always had a reputation as the most moderate of the southern states. After the *Brown* decisions, Governor Luther Hodges led the way in seeing that blacks were admitted to white schools. But even he could only ensure the enrollment of a few "token Negroes." And, with regard to them, *moderation* remained only a relative term. The first black girl to enter one white Charlotte school was met by a jeering mob. "I started walking toward the school, a lot of people were just pushing and shoving and calling me names and throwing things, spitting on me and saying . . . 'Nigger, go back to Africa.' "[25]

By 1965, when the *Swann* case began, the situation had improved, but not by that much. The violence had long since abated and more blacks were attending white schools. However, the number was still depressingly small. A mere 2 percent of the black students in Charlotte—490 out of 20,000—were in schools with whites. More than 80 percent of the 490 were in one school with 7 white pupils; the remainder were distributed among 7 of the 103 schools in the school district. Virtually all blacks in Charlotte were still attending all-black schools.[26]

One important step had, nevertheless, been taken, though it was not at all intended to aid desegregation. In 1960, county and city leaders joined to establish the Charlotte-Mecklenburg School District by merging the Charlotte City Public Schools with the Mecklenburg County Public Schools. The merger's purpose had nothing to do with desegregation; the aim was to create a metropolitan school district that could be operated more efficiently than the separate urban and rural districts that had previously existed. But the new district was also one from which the dual school system (with separate black and white schools) could more readily be eliminated. It may be doubted that Judge McMillan's desegregation orders in the *Swann* case could have been effective if they had been able to reach only Charlotte's urban schools. With the case encompassing the entire county, as well as the urban center, the judge could order a plan which, as we shall see, paired white and black schools throughout the metropolitan district.

Swann Under Way

In 1961, when Judge McMillan had made his North Carolina Law Review Association speech against school busing, seated in the audience was Julius LeVonne Chambers, then a law student at Chapel Hill and the first black editor-in-chief of the *North Carolina Law Review*—a highly significant accomplishment for a black in a major southern law school. Like McMillan, Chambers had grown up in rural North Carolina, about forty miles east of Charlotte, but on the other side of the color line at a time when Jim Crow was still dominant in the South. Chambers, too, had to travel miles to high school, but he was bused past a far better white school close to his home and he noted the irony daily.

Chambers's independence was shown at an early age. As a teenager he decided he wanted a third name. Born LeVonne Chambers, he became Julius LeVonne Chambers. There was no particular reason for choosing the name Julius; it just had a good sound to him. "He started out to be a pretty good auto mechanic," Chambers's father said recently. "I thought he was going to stay with that." But Chambers changed his career goal when his father, who ran a service station, was unable to obtain an attorney to help him collect a bill for a truck repair he had done for a white person. "The guy who owed the debt was a pretty powerful figure in the community, and lawyers there didn't want to represent a black person," says Chambers in his quiet but firm voice. "My father brought the truck home, and the man came and drove it off." This incident, when he was sixteen, made Chambers want to become a lawyer.[27]

At North Carolina Law School, Chambers ranked first in a class of one hundred and won the coveted position of editor-in-chief of the *Law Review*. At that time Chapel Hill was still a difficult place for blacks. A fellow black law student remembers watching the TV news with Chambers during the period of arrests for sit-ins and other demonstrations. "We sat there and burned slowly," he says. Chambers himself, however, remembers only his concentration on his work. "I studied all the time."[28] Even McMillan's speech did not evoke any real protest. At that time, Chambers had no idea of his future role in vindicating civil rights.

Chambers is of medium height, somewhat stocky, with short hair now flecked with traces of gray. He has always dressed conservatively and is a voracious reader. After his graduation from North Carolina Law School and graduate work at Columbia, where he received an LL.M. degree in 1963, Chambers became the first intern at the NAACP Legal Defense Fund (he is now the fund's director-counsel). A year later he returned to North Carolina and set up an LDF-assisted practice in Charlotte. In return for financial aid in starting his practice, he agreed to assist in cases the LDF wanted brought. Over forty other young black lawyers have similarly been assisted by the LDF over the years.

Civil rights iconography has it that desegregation cases begin with the outrage of the black parent whose child is turned away from a white school that is typically closer and educationally superior to the one he is required to attend. According to a 1984 report, this was true of the *Swann* case as well: "In 1964, James, the six-year-old son of Vera and Darius Swann, was denied admission to a predominantly white school near his home on the campus of Johnson C. Smith University, where his father was a theology professor. The landmark civil rights case bearing his name was [then] filed."[29]

The reality is that neither young Swann nor his parents had much to do with initiating the case that has made their name a civil rights byword. According to Chambers, it was the decision of the Charlotte-Mecklenburg School Board to deal with its desegregation obligations by closing several all-black schools that triggered the filing of an action. As Chambers told me, "We in the city and county had been looking at the steps the board was taking to desegregate the schools and were concerned with the limited progress that was being made." In particular, Chambers noted, "We were upset with the board's proposal simply to close several black schools in the county rather than take some steps that would effectively desegregate the schools. Working with the NAACP and several local black groups, we decided to file suit."

Chambers and the others were able to obtain the consent of ten black families to bring suit on behalf of their twenty-five children and any other black children similarly affected. The Swanns were among these parents. Traditionally, a case's name is taken from the first plaintiff listed in the complaint. "There was some concern," Chambers has said, about finding the appropriate "lead plaintiff because that person would be the one most frequently referred to." The Swann child was chosen because his father had a secure university position and hence was less vulnerable to economic pressure and other forms of local retaliation than the other plaintiffs. James E. Swann thus became the first named plaintiff.

Yet, whether or not the Swanns were responsible for conceiving the litigation that started on January 19, 1965, it is as the *Swann* case that this litigation has gone down in constitutional history. It is with the Swann name that busing as a remedy for school desegregation is permanently associated, even though the man most responsible for the case was the attorney for the plaintiffs, Julius L. Chambers.

When Chambers brought the *Swann* action, we saw, only 490 of the more than 20,000 black schoolchildren in the school district were attending schools with white children.[30] "We filed a lawsuit in 1965," Chambers explains, "with the objective of getting the court to impose an affirmative obligation on the school board to disestablish the segregated system that it had created, and to effect as much desegregation in the system as possible."[31]

As already noted, the immediate catalyst for the *Swann* lawsuit was the Charlotte School Board's plan to close some black schools. The plan also contained a provision for attendance zones for most of the schools in the district, and a "freedom of choice" provision under which students could transfer to any school in the district provided they could furnish their own transportation and the school was not already filled to capacity. Plaintiffs challenged the plan on the ground that the closing of the black schools would place the burden of desegregation on the black pupils and that other features of the plan would only perpetuate segregation.

In 1965, *Swann v. Charlotte-Mecklenburg Board of Education* went before the district court. The judge who heard the case, J. Braxton Craven, rejected plaintiff's challenge and approved the school board's plan. In response to the claim that attendance zones should be changed "to increase mixing of the races," Judge Craven declared, "I know of no such duty upon either the School Board or the District Court."[32] A year later, the Court of Appeals for the Fourth Circuit affirmed the lower-court decision, stating "that there is no constitutional requirement that [the board] act with the conscious purpose of achieving the maximum mixture of the races in the school population."[33] At this point, Chambers decided not to go to the Supreme Court. He feared that, under the precedents at that time, the Court would only affirm the court of appeals and thus lend added weight to that tribunal's restrictive decision.

But in the 1968 and 1969 Supreme Court terms the legal situation completely changed. In *Green* v. *County School Board*[34] (to be discussed in Chapter 3), the Court ruled that "freedom of choice" plans do not satisfy the requirement of *Brown I* and *II*. In doing so, the Court appeared to repudiate the limited responsibility for desegregation found by Judge Craven and the court of appeals. According to the *Green* opinion, school boards were "clearly charged with the affirmative duty to take whatever steps might be necessary to convert to a unitary system in which racial discrimination would be eliminated root and branch." This affirmative duty to desegregate carried with it the "burden . . . to come forward with a plan that promises realistically to work, and promises realistically to work *now*."[35]

As we will see in Chapter 3, the *Green* decision was one of the most important school desegregation decisions by the Warren Court. The next year, under its new head, Chief Justice Warren E. Burger, the Court decided *Alexander* v. *Holmes County Board of Education*[36] (to be discussed in Chapter 4). In a brief order, reaffirming what had been said in *Green,* it reversed a court of appeals order granting a three-month delay to the Holmes County Board of Education in desegregating its schools. Instead, the Court directed the court of appeals to order "that each of the school districts here involved may no longer operate a dual school system based on race or color, and directing that they begin immediately to operate as unitary school systems within which no person is to be effectively excluded from any school because of race or color."[37]

During the 1968–69 school year, the schools in Charlotte continued to operate under the plan that had been approved by the federal courts in 1965 and 1966. This meant that, despite *Green,* some 14,000 of the 24,000 black students in the district still attended schools that were all or predominantly black. Most of the 24,000 had no white teachers.

The *Green* decision cast serious doubt on the validity of the plan under which Charlotte's schools were still operating. The court of appeals opinion upholding the plan had said that there was no affirmative duty to secure racially mixed schools. *Green* directly repudiated a similarly restrictive approach, asserting, as we saw, that the school board did have the affirmative duty to convert to a unitary system and to do so by plans that promised realistically to accomplish that end and to do so *now*. "The duty now appears as not simply a negative duty to refrain from active legal racial discrimination, but a duty to act positively to fashion affirmatively a school system as free as possible from the lasting effects of such historical *apartheid*."[38]

The *Green* decision led Chambers and his associates to reopen the *Swann* case. At that time, he says, we had an interest in developing a case that would allow us to present the issue of whether transportation would have to be required in these cases. Richmond, Virginia, was considered, the Mobile school case was considered, Charlotte was considered along with two or three others. And we began in 1968 with a motion for further relief in which we wanted to present the issue to the court whether the school district in complying with *Brown* would have to take affirmative steps to desegregate and would have to use transportation if that was necessary to effect complete desegregation in the system.

On September 6, 1968, the *Swann* plaintiffs filed a "Motion for Further Relief" in the federal district court in Charlotte. The motion sought greater speed in desegregation of the Charlotte schools and requested elimination of alleged racial inequalities. The motion came before Judge James B. McMillan. This was the first important case assigned to McMillan upon his appointment to the federal bench.

Trial and Findings

With the *Swann* case, says Judge McMillan, "my factual education began" into the realities of the school segregation that still existed in Charlotte fifteen years after the *Brown* decision[39] had ruled segregation unconstitutional. The judge's education took place during the trial on the *Swann* motion. Evidence was taken at length by McMillan during much of March 1969, as well as during June of that year. As Chambers and his colleagues presented the evidence in support of the motion, what the judge was later to term "my uninformed . . . Olympian certainty" about the Charlotte schools as a model of desegregation, in which "[m]any black children were going to 'white' schools," began to give way. When the case came to him, notes McMillan, "I first said, 'What's wrong in Charlotte?' " As the evidence in the case accumulated, he began to realize the extent to which segregation still existed in the Charlotte school system. The result was, as McMillan explains it, "I got in the position that I had to act on something that was based on fact and law rather than feelings."[40]

What were the facts, as revealed by the evidence, that led Judge McMillan to conclude that drastic steps were needed if Charlotte-Mecklenburg was to be made into a unitary school system in accordance with the command of the *Green* case?

Judge McMillan has said that, "though opinions differed as to remedies, the findings of *fact* on which my various orders were made were never in serious conflict."[41] These findings were summarized in the Supreme Court's *Swann* opinion as follows:

The Charlotte-Mecklenburg school system, the 43d largest in the Nation, encompasses the city of Charlotte and surrounding Mecklenburg County, North Carolina. The area is large—550 square miles—spanning roughly 22 miles east-west and 36 miles north-south. During the 1968–1969 school year the system served more than 84,000 pupils in 107 schools. Approximately 71% of the pupils were found to be white and 29% Negro. As of June 1969 there were approximately 24,000 Negro students in the system, of whom 21,000 attended schools within the city of Charlotte. Two-thirds of those 21,000—approximately 14,000 Negro students—attended 21 schools which were either totally Negro or more than 99% Negro.

In view of these facts, "All parties now agree that in 1969 the system fell short of achieving the unitary school system that those cases [i.e., *Green* and its companion cases] require."[42]

The school board argued that it had made a good-faith effort to desegregate. As one of the board's attorneys recently put it, "We truly felt that this school system had done what complied with the constitutional mandates." Blacks had freedom of choice, but most wanted to attend schools in their own neighborhood. Judge McMillan rejected this argument. His own findings were that the city schools in the district "are still largely segregated." The evidence showed that "most black students attend totally or almost totally segregated schools" and "most white students attend largely or completely segregated schools."[43] More than that, the judge found that the segregation that still existed in the school system in Charlotte "is caused by government action" and "has its roots in law."[44]

The result of all this, according to McMillan, was a patent violation of the *Green* requirement of immediate desegregation that led to inferior education for black students. The latter conclusion was supported by the testimony of three expert witnesses from Rhode Island College, who agreed that a racial mix in which black students heavily preponderated tended to retard the progress of the whole group. The experts also submitted a 55-page report which outlined several plans by which desegregation in the Charlotte schools might be accomplished.

Judge McMillan recognized that, of the three experts in education who testified for plaintiffs, "None was as familiar with the local situation as the local Board and school administrators."[45] It should, however, be recognized that there were practical problems in securing experts who were closer to the case than those who did testify.

Chambers later commented on this problem. "Educators in the area who we felt would have been reputable and persuasive would not dare get involved in litigation as witnesses for plaintiffs at this period. It would have meant loss of jobs, loss of friends and so one had to go to other areas to find educational experts to assist."

As for Judge McMillan's transformation, one of the attorneys for the school board asserts that "as we got into the case, I was surprised to find that his attitude became quickly an adversary one to us and we felt that we were dealing more with meeting the judge's arguments than . . . the opposition's arguments."[46] But McMillan says that he was also surprised because he had expected his decision to come out differently when he started the case. At that time, as already noted, he had asked, "What's wrong in Charlotte?" "I set the case for hearing reluctantly. I heard it reluctantly, at first unbelievingly."[47]

Now the evidence had given him a more accurate picture of the true situation on the education of blacks and how this situation had been brought about by government action in the school district. McMillan had no doubt that the Charlotte system violated the requirement laid down by the Supreme Court in the *Green* case. Perhaps the earlier decisions upholding the school board's plan were correct at the time they were decided, but the "rules of the game have changed and the methods and philosophies which in good faith the Board has followed are no longer adequate." Since the board's plan was approved, "the law has moved from an attitude barring discrimination to an attitude requiring active desegregation. The actions of School Boards and district courts must now be judged under *Green v. New Kent County* rather than under the milder lash of *Brown v. Board of Education.*"[48]

Having reached the conclusion that the way the Charlotte school system was operated violated the *Green* requirement—that its nondiscrimination approach was "no longer adequate to complete the job which the courts now say must be done 'now' "[49]—what could Judge McMillan do except order remedial action? "After the facts began to be assembled," the judge recalls, "and I began to deal in terms of facts and information instead of in terms of my natural-born raising, I began to realize and finally advised the parties that something should be done."[50]

First Desegregation Orders

On April 23, 1969, less than a month after he had finished hearing the evidence, Judge McMillan issued his first opinion and order in the case. The opinion summarized the evidence and concluded categorically that the Charlotte school system was still segregated: "The relatively complete extent of the segregation of the schools in this system is demonstrated by study of the . . . statistics." Consequently, the plan under which the school operated must now be ruled invalid. Under *Green* it was the school boards who "are now clearly charged with the affirmative duty to desegregate schools 'now' by positive measures." The board was therefore directed to submit by May 15 "a plan for effective desegregation of pupil population, to be predominantly effective in the fall of 1969 and to be completed by the fall of 1970." The board was to be free to use any of the different methods which had been advanced, including busing and changing of school zones. The object was to be a "plan . . . for the effective operation of the schools in a desegregated atmosphere."[51]

Judge McMillan's thought was to proceed as slowly as he could on the remedial aspect of the case.[52] Instead of imposing a desegregation plan upon the school board, he first tried "the compromise" of directing the board to prepare its own plan. What happened was described by McMillan in his 1981 testimony at a Senate hearing: "I asked the School Board to make those changes on their own and . . . they declined. The invitation was repeated several times by formal order throughout 1969, and a consistent five to four majority of the board declined the invitation."[53]

Julius Chambers thought that McMillan's initial remedial approach was "contrary to the *Alexander* short opinion,"[54] which, as Chambers saw it, "required immediate steps to eliminate [segregation] and in fact we challenged his refusal to order desegregation in the middle of the school year." But, Chambers concedes, McMillan "wanted to come to a decision which would be accepted by both sides if possible. . . . He later found out that it was a futile objective."

In June 1969, Judge McMillan denied a motion filed by plaintiffs for a contempt citation against the school board members. Though he recognized that the board had not come up with a satisfactory plan as he had ordered, "on an issue of such significance, the amount of foot-dragging which has taken place, up to now at least, should not be considered as contempt of court."[55]

McMillan did, however, again order the board to prepare and submit "a positive plan for desegregation of the pupils of the Charlotte-Mecklenburg school system"[56]—this time by August 16, 1969. Then, on July 29, the board came forward with a weak plan that McMillan approved on an interim basis, though "only . . . with great reluctance." He directed the board to file another plan by November 17 that would provide for "the complete desegregation of the system to the maximum extent possible" for the year 1970–71.[57]

In November the board moved for an extension of time until February 1970. Judge McMillan denied the motion on November 7.[58] A week earlier, in the *Alexander* case,[59] the Supreme Court had peremptorily reversed an order granting a three-month delay in school desegregation.

McMillan ruled that the *Alexander* "prohibition against extension of time"[60] was binding and barred his granting defendant's motion. The board then submitted a November 17 plan, which was found completely unsatisfactory by McMillan. According to the judge, "If the courts should accept the defendants' contention that all they have to do is re-draw attendance lines and allow a type of freedom of choice, two-thirds or more of the black children in Mecklenburg County would be relegated permanently to this kind of separate but *un*equal education."[61]

By this time, Judge McMillan had come to recognize that the school board itself would not come forward with a satisfactory desegregation plan. If the segregation that still existed in the Charlotte-Mecklenburg schools was to be ended, it would have to be by a plan ordered by the judge himself. On December 1, 1969, McMillan issued an order disapproving the board's November 17 plan. The order further provided, "A consultant will be designated by the court to prepare immediately plans and recommendations to the court for desegregation of the schools."[62] Charlotte-Mecklenburg was now on notice that a court-appointed expert would prepare a desegregation plan whose terms would be enforced by the court itself.

McMillan's Busing Order

During the *Swann* hearing in June 1969, Dr. John A. Finger, Jr., of Rhode Island College, had testified as an expert witness for the plaintiffs. At that time he described in detail a plan formulated by him by which the Charlotte schools could be desegregated. In his June 1969 order, Judge McMillan stated that the local school staff considered such a plan feasible. The judge went on to say that "the local school administrative staff are also better equipped than Dr. Finger, a 'visiting fireman,' to work out and put into effect a plan of this sort."[63]

Despite his earlier characterization of Dr. Finger as an outsider, and moreover one who had testified as a witness for plaintiffs, the judge now appointed Finger as his consultant to "advise the court how the schools could be desegregated."[64] The Finger appointment would later lead the court of appeals, on review of Judge McMillan's order, to state, "We caution . . . that when a court needs an expert, it should avoid appointing a person who has appeared as a witness for one of the parties."[65] This animadversion was almost repeated by the Supreme Court. The early drafts of Chief Justice Burger's *Swann* opinion of the Court would also contain a "caveat" on the appointment of an expert who had testified for one of the parties: "Expert witnesses with an interest in sustaining their own plans and positions are placed in an awkward position at best."[66] This passage was omitted from the final Burger *Swann* opinion.

When later asked why he had appointed a witness for plaintiffs as his consultant, Judge McMillan answered, paralleling Julius Chambers's statement about why he had felt constrained to go to Rhode Island for Dr. Finger, that it was not possible for him to secure an expert from the area who was willing to be appointed. This, he said, tilted the balance against his normal reluctance to select a consultant who had testified for one of the parties.

After Dr. Finger's appointment, the school board submitted yet another desegregation plan in February 1970. This plan was also rejected by the judge as "still inadequate because it leaves half the black elementary students still attending black schools."[67] and "promises to provide little or no transportation in aid of desegregation."[68]

At the same time Dr. Finger, who was working on his own plan, presented his plan to the court. The Finger plan had, in fact, been prepared with the substantial, though unofficial, help of a number of members of the school board staff: "The school staff worked out the details of this plan and are familiar with it," Dr. Finger would tell the Court.[69] The aid given had been unofficial and, indeed, kept secret at the time; staff members known to have cooperated with Finger would undoubtedly have been disciplined by the school board. But the help was real and of great importance in making the Finger plan a workable one. As Judge McMillan later described it, "Finger was the catalyst through which the board staff could work with the court."

The Finger desegregation plan was far-reaching. It provided for the desegregation of all 105 Charlotte-Mecklenburg schools. It did so by pairing and clustering groups of schools so that every white school district would have a corresponding satellite district in a black community. Thus, "the proposed attendance zones for the high schools were typically shaped like wedges of a pie, extending outward from the center of the city to the suburban and rural areas of the county in order to afford residents of the center city area access to outlying schools."[70] Inner-city black students were assigned to outlying, predominantly white junior and senior high schools, thereby substantially desegregating all the system's secondary schools.[71]

The Finger plan desegregated the elementary schools "by the technique of grouping two or three outlying schools with one black inner city school; by transporting black students from grades one through four to the outlying white schools; and by transporting white students from the fifth and sixth grades from the outlying white schools to the inner city black school."[72]

Free transportation was to be provided for all students who were assigned to schools outside their neighborhoods. The result would be the busing of about 10,000 students solely for the purpose of desegregation[73]—about one-fourth of the children being transported in Charlotte-Mecklenburg.[74]

"Jack," Judge McMillan recalls telling Dr. Finger after seeing his plan, "this is political dynamite and will cause a real commotion. But let's go ahead." On February 5, 1970, McMillan issued an opinion and order adopting the Finger desegregation plan. The board was directed to desegregate the schools in Charlotte-Mecklenburg in accordance with Dr. Finger's plan[75] and to provide "transportation . . . on a uniform nonracial basis to all children whose reassignment to any school"[76] was brought about by the plan. The judge further ordered that desegregation under the Finger plan was to be accomplished by April 1, 1970, for elementary schools, and by May 4 for junior high and most high school students (only high school seniors might remain in their present schools until the end of the school year).

"This will raise a lot of hell," McMillan told his law clerk when he had decided to order the Finger plan. If anything, the judge's prediction understated the storm of controversy that would arise over his order. Most of the opposition to the order focused on the busing requirement imposed by the judge. McMillan himself still asserts that the busing issue was an "absolutely false trail." He caustically refers to the "crocodile tears for young children that were never shed for those riding up to fifty miles or more per day before the case." He points out that three-fifths of the public school students in North Carolina rode school buses every day and no one objected until busing was used as a desegregation tool. Before *Swann*, indeed, the school bus was "a source of pride for Americans especially in North Carolina, which during the 1950s labeled itself as 'the school-busingest state in the Union.' "[77]

The same was true, Judge McMillan maintained, of the "neighborhood school" theory upon which opponents of the *Swann* decision relied. "When racial segregation was required by law, nobody evoked the neighborhood school theory to *permit* black children to attend white schools close to where they lived." Besides, even if there were some validity to the neighborhood school theory, it had to give way to the desegregation requirement. "The neighborhood school theory has no standing to override the Constitution."[78]

But as an *Atlantic Monthly* article pointed out in 1972, it was more than the transportation requirement involved in a busing order such as that issued by Judge McMillan that was so disturbing: "Busing, the very term, is actually a shorthand symbol for the redistribution of a city's student population, leaving no school segregated and, in certain localities, each school with an enrollment approximating the white-to-black ratio in the population at large—about seven to three in Charlotte."[79]

In his December 1, 1969, opinion and order (the one in which he announced the appointment of Dr. Finger), McMillan had stated, "the court will start with the thought . . . that efforts should be made to reach a 71–29 ratio in the various schools so that there will be no basis for contending that one school is racially different from the others."[80] In addition, the judge declared "[t]hat pupils of all grades [should] be assigned in such a way that as nearly as practicable the various schools at various grade levels have about the same proportion of black and white students."[81]

This language led, as we shall see, to a difference of opinion among the Supreme Court Justices on whether the judge had imposed a fixed mathematical racial balance requiring all the Charlotte-Mecklenburg schools to reflect the racial composition of the school system as a whole. Ultimately the Supreme Court concluded that McMillan did not "require, as a matter of substantive constitutional right, any particular degree of racial balance or mixing." Instead, "the use made of mathematical ratios was no more than a starting point in the process of shaping a remedy."[82]

McMillan himself strongly denied that he had "ruled . . . that 'racial balance' is required under the Constitution."[83] His February 5, 1970, desegregation order appears to bear him out on this point. It expressly contemplated wide variations in permissible school populations. Thus, the order approved the Finger plan for the elementary schools (the most controversial portion of the court consultant's plan), which provided for schools with pupil populations ranging from 3 percent black at one extreme to 41 percent at the other.[84] Of the other elementary schools under the Finger plan, 4 would have fewer than 20 percent blacks, 21 would have between 21 and 29 percent, and 23 would have between 31 and 38 percent black pupils.[85] Well might McMillan say, "This is not racial balance but racial diversity. The purpose is not some fictitious 'mix,' but the compliance of this school system with the Constitution by eliminating the racial characteristics of its schools."[86]

Though the McMillan order was not as rigid regarding racial balance as its opponents contended, this fine point scarcely appeased the bitterness of those who read or heard of the decision in bits and pieces. McMillan's decision, says one observer, "set off a sociological earthquake."[87] According to a *Washington Post* story, the judge's "far-reaching bus-and-balance order . . . triggered widespread criticism and abuse from the white community, anonymously telephoned death threats, picketing on his lawn and a need for police guards."[88]

After the order, McMillan was hanged in effigy. Sign-bearing crowds demonstrated at the U.S. courthouse, in front of the judge's wood-paneled brick house, and at the *Charlotte Observer,* a newspaper that had supported busing. Ministers denounced busing from the pulpit, and the other local paper, the *Charlotte News,* condemned McMillan in its editorials. Groups of all sizes, from the Queen City Jaycees to the Classroom Teachers Association, came out against busing. The Charlotte streets blossomed with thousands of red, white, and blue "No Forced Busing" bumper stickers, and the Concerned Parents Association, a new organization formed to fight McMillan's order, collected 80,000 signatures on antibusing petitions. The bitterness in Charlotte is shown by the comment of a veteran at an antibusing rally in a school auditorium: "I served in Korea, I served in Vietnam, and I'll serve in Charlotte if I need to."[89]

McMillan would have had to be extremely thick-skinned not to have been affected by the campaign against him. Speaking of the judge, Julius Chambers recalls, "I know that during the period he was receiving all kinds of threatening phone calls. He had to have a personal guard at his home. His wife was receiving threatening calls and he was ostracized here in the community. He'd go to play golf and somebody would find that he didn't want to play or couldn't play with him. And I think it's even affecting him today."

Chambers himself suffered even more directly. In February 1971, his law office was firebombed by an arsonist's midnight attack;[90] not long after he had originally brought the *Swann* case, his car was dynamited and, some months later, his home was hit by what the police believed was dynamite.[91] Chambers's stoicism in the face of these attacks became almost legendary in Charlotte. When his auto was bombed on a public street, Chambers was speaking to a group of blacks inside a nearby church. "Julius didn't get excited at all," recalls an attorney who was present at the incident. "He went and took a look at his car, and went back and finished his speech."[92]

Court of Appeals Reversal

The public pressure and acts of violence received the headlines, but they had little effect on the resolution of the Charlotte-Mecklenburg desegregation issue. That issue was decided in a legal, not a political, arena. At a Senate hearing in 1981, Judge McMillan was asked whether it was not necessary to have "some degree of responsiveness to public sentiment" on school busing. He replied that "a judge . . . cannot go by public opinion. That is the problem Senator East has got—trying to cope with public opinion. I cannot cope with public opinion in dealing with the rights of man under the Constitution, the law of the land. I cannot go by a vote of the neighbors or the electorate at large."[93]

On the bench, McMillan had expressed the same view: "To yield to public clamor, however, is to corrupt the judicial process and to turn the effective operation of courts over to political activism and to the temporary local opinion makers. This a court must not do."[94] Or, as the judge stated it in simpler terms in another of his Swann opinions, "A correspondent who signs 'Puzzled' inquires: 'If the whites don't want it and the blacks don't want it, why do we have to have it?' The answer is, the Constitution of the United States."[95]

After Judge McMillan had issued his February 5, 1970 order adopting the Finger desegregation plan, with its provision for school busing, the only legal method available to block its enforcement was an appeal to the U.S. Court of Appeals for the Fourth Circuit. The Charlotte-Mecklenburg School District did just that. Implementation of the Finger plan was partially stayed by the appellate court on March 5, pending a hearing of the appeal. The Supreme Court refused to disturb the fourth circuit's stay order.[96]

A federal court of appeals normally sits in panels of three. In important cases, however, the court may sit en banc, with all of the judges on the court participating in the case. This is what happened in the Swann case. Six of the seven fourth circuit judges were on the bench (the seventh had disqualified himself because of prior participation in the case)[97] when the Swann appeal was argued on April 9. Like Judge McMillan, all the fourth circuit judges were from states with strong segregationist histories: Three were from Virginia, two from Maryland, and one each from North and South Carolina. At the center of the fourth circuit bench sat Chief Judge Clement F. Haynsworth, whose nomination to the Supreme Court by President Nixon had been turned down by the Senate.

After the argument it was quickly apparent that the court of appeals was sharply divided on the case. Two judges, Butzner and Boreman, agreed with the McMillan plan for senior and junior high school students but thought that he went too far in his order for extensive busing of elementary school students. Judge Butzner circulated an opinion to that effect. It adopted a "reasonableness" test by which to judge such desegregation orders. "Viewing the plan the district court approved for junior and senior high schools against these principles and the background of national, state, and local transportation policies, we conclude that it provides a reasonable way of eliminating all segregation in these schools." The same, according to the Butzner opinion, was not true of the elementary schools. Bearing in mind the age of the pupils there, it was not reasonable to require the extensive busing that the district court had ordered for them. The district court judgment was therefore vacated and the case remanded "for further consideration of the assignment of pupils attending elementary schools."[98]

On May 15, Judge Haynsworth wrote to Judge Butzner that he concurred in the opinion. At the same time, he indicated that while he was willing to go along generally with busing junior and senior high school students, he was inclined "to file a separate opinion. The principal thought I have in mind is an extension of your notion of granting reasonable flexibility." Haynsworth referred to McMillan's requirement that an additional 300 blacks be assigned to one white high school. He thought "we should affirm Judge McMillan's requirement . . . but I am gravely concerned about our treating school children as ciphers, indistinguishable except for race and without individual personalities." If the 300 and their parents "stoutly object" and experience "shows that successful adjustments for them are improbable . . . I would want the Board and the District Court to know that so highly undesirable an educational situation need not forever be continued with the three hundred serving as sacrificial lambs to an inflexible demand for racial mixing."[99]

Haynsworth never issued his separate opinion (presumably because of the court's fragmentation), but simply joined Judge Butzner. The Haynsworth joinder gave the Butzner opinion the support of three of the fourth circuit judges. However, Judges Sobeloff and Winter disagreed with the reversal of Judge McMillan on the elementary schools. They wrote opinions dissenting "from the failure to affirm the portion of the order pertaining to the elementary schools." Instead, they "would affirm the order of the district court in its entirety."[100]

The sixth court of appeals judge, Judge Bryan, also dissented. But he did so because he thought that the busing order should be reversed in its entirety. As Bryan saw it, in affirming the senior and junior high school parts of McMillan's order, "The Court commands the Charlotte-Mecklenburg Board of Education to provide busing of pupils to its public schools for 'achieving *integration.'*" But the objective was to achieve "racial *balance*" and that was not required by the Constitution. "I would not, as the majority does, lay upon Charlotte-Mecklenburg this so doubtfully Constitutional ukase."[101]

When he sent his opinion to Judge Butzner on May 12, 1970, Judge Bryan wrote that his dissent was "Like a boy throwing a stone at Napoleon's army." It should have had a much greater effect than that, for it left the court divided three-two-one, with no clear majority for the Butzner opinion. When, despite this, Judge Butzner drafted a mandate giving effect to his opinion, Judges Sobeloff and Winter joined in a "protest," asserting, "We do not agree that the proposed mandate is warranted by the opinions in this case."[102]

Sobeloff indicated that he realized that the three-two-one division and the probable public dispute on the judgment would scarcely enhance the court's reputation. "I half hope and predict," he wrote to Winter, "that Judge Bryan, when he fully realizes the situation, will join the three [votes to modify] and make our point academic."[103] Had Bryan, who wanted to reverse McMillan's decision in its entirety, refrained from joining the Butzner opinion to uphold it in part, the lack of a majority would have left the lower-court decision in effect in its entirety.

In their "protest" filed with the clerk to be attached to the mandate drafted by Butzner, Sobeloff and Winter stated that the Bryan opinion "express[es] disagreement with busing of pupils to public schools to achieve a unitary system."[104] Bryan objected to this and wrote to Winter, "In the first paragraph of what I wrote I attemped to distinguish between the busing of pupils to achieve *integration* and busing to achieve *balance*. Only the latter did I attack."[105]

Bryan then decided to resolve the matter by expressly joining the Butzner-drafted mandate. The judgment accompanying the opinions issued by the court of appeals could consequently read, "Judge Bryan joins Haynsworth, C.J., and Boreman, J., in voting to vacate the judgment of the District Court, and to remand the case in accordance with the opinion written by Butzner, J. He does so for the sake of creating a clear majority for the decision to remand."[106]

A clear majority thus joined the fourth circuit decision to vacate Judge McMillan's order respecting elementary schools, and affirming only on the secondary school plans. McMillan was noted in the press as having "felt the rug had been pulled [out] from under him." On the other hand, a leader of the Charlotte protesters was quoted as saying that "it does show public opinion of the masses does make a difference, politically and judicially."[107]

Soon after the court of appeals decision, however, *Swann* was once more on its way—this time to the Supreme Court after the plaintiffs petitioned that tribunal for review. The highest Court has always been noted for its insulation from political and other pressures. All the same, Mr. Dooley has not been alone in noting that "th' supreme court follows th' iliction returns."[108] Not long before Judge McMillan issued his first *Swann* order, Richard M. Nixon had been elected to the presidency. In March 1970, Nixon had issued a policy statement condemning "forced" busing and supporting neighborhood schools.[109] Now the new Chief Justice, whom he had appointed, was to preside over the case that would help to determine whether the Dooley dictum would apply to the Burger Court.

2

The Burger Court and
the Justices

The Supreme Court that would decide the *Swann* case had just under-
gone a dramatic personnel change. From October 5, 1953 to June 23,
1969, Earl Warren had presided over the Court. During those years he
had led a judicial revolution that had transformed both the law and the
society it served. When Warren became Chief Justice, Jim Crow was the
dominant feature in southern life, rural America held the reins of government, and the third degree and abusive police practices were still the
hallmarks of the criminal law. By the time Warren retired, racial discrimination had been placed beyond the constitutional pale; Court-ordered
reapportionment to approximate the one-man, one-vote ideal had transferred political dominance to the cities and suburbs, where most Americans lived; and protection of the rights of criminal defendants had become a prime judicial priority.

"America had changed irrevocably—and the Warren Court had been
midwife to that change, if not its sire."[1] In expanding civil rights, extending the franchise, reinforcing freedoms of speech, assembly, and religion,
and defining the limits of police authority, the Warren Court had no
equal in American history.

Now the Warren era had ended. The strongest Chief Justice since
John Marshall had stepped down and been succeeded by a judge who
was still an unknown quantity. It was this Court, headed by a new Chief
Justice, that would make the decision in the *Swann* school busing case.

The Chief Justice's Leadership Position

The Supreme Court is an institution whose collegiate nature is underscored by the custom the Justices have of calling each other "Brethren."

But as Brethren the Justices can only be guided, not directed. As Justice Frankfurter once stated in a letter to Chief Justice Vinson, "good feeling in the Court, as in a family, is produced by accommodation, not by authority—whether the authority of a parent or a vote."[2]

The Court "family" is composed of nine individuals who constantly bear out the truism that "judges are only men."[3] "To be sure," Frankfurter once wrote to Justice Reed, "the Court is an institution, but individuals, with all their diversities of endowment, experience and outlook determine its actions. The history of the Supreme Court is not the history of an abstraction, but the analysis of individuals acting as a Court who make decisions and lay down doctrines."[4]

In many ways, the individual Justices operate, as Justice Powell has said, like "nine small independent law firms."[5] In such a Court, it is most difficult for a Chief Justice to assert a *formal* leadership role. While it may be the custom to identify a sitting high Court by the name of its Chief, one who looks only to the bare legal powers of the Chief Justice will find it hard to understand this underscoring of his preeminence. Aside from his designation as Chief of the Court and the attribution of a slightly higher salary, his position is not superior to that of his colleagues—and it certainly is not legally superior.[6]

Understandably, the Justices themselves have always been sensitive to claims that the Chief Justice has greater power than the others. "It is vitally important," asserted Justice Frankfurter in a 1956 letter to Justice Burton, "to remember what Holmes said about the office: 'Of course, the position of the Chief Justice differs from that of the other Justices only on the administrative side.' He is not the head of a Department; not even a quarterback."[7] Two years later, Frankfurter wrote Justice Brennan "that any encouragement in a Chief Justice that he is the boss . . . must be rigorously resisted. . . . I, for my part, will discharge what I regard as a post of trusteeship, not least in keeping the Chief Justice in his place, as long as I am around."[8]

Notwithstanding the executive strength Earl Warren brought to his position, even he quickly realized that he was not going to be able to deal with the Justices in the same way he had directed matters as chief executive of California. "I think," Justice Stewart told me, "he came to realize very early, certainly long before I came here [1958], that this group of nine rather prima donnaish people could not be led, could not be told, in the way the governor of California can tell a subordinate, do this or do that."

A study of the Chief Justiceship should not, however, be approached only in a formalistic sense. Starting with Marshall, the greatest of the Chief Justices have known how to make the most of the potential extra-legal power inherent in their position. The Chief Justice may be only *primus inter pares,* but he is *primus.* Somebody has to preside over a body of nine, and it is the Chief Justice who does preside, both in open court and in the even more important work of deciding cases in the conference chamber.[9] It is the Chief Justice who directs the business of the Court. He controls the discussion in conference; his is the prerogative to call and discuss cases before the other Justices speak.

The most important work of the Court is done in private, particularly in the conference sessions that take place after cases have been argued before the Justices. This is no new tradition; referring to the Justices of the Taney Court, Justice Campbell once said, "Their most arduous and responsible duty is in the conference."[10] It is still true. It is in the conference that the Court decides the cases that have come before it.

The primary role at the conference is exercised by the Chief Justice. It is his function to lead the discussion. He starts the conference on each case by stating the facts and issues involved. He then discusses the issues and tells how he would vote. Only after the Chief Justice has finished his presentation do the other Justices have an opportunity to state their views, which they do in order of descending seniority.

The manner in which he leads the conference is the key to much of a Chief Justice's effectiveness. His presentation fixes the theme for the discussion that follows and, if skillfully done, is a major force in leading to the decision he favors.

The two strongest leaders of the conference during this century were Chief Justices Charles Evans Hughes and Earl Warren. Most students of the Court rank Hughes as the most efficient conference manager the Justices have ever had. He imposed a tight schedule on case discussions. Long-winded and irrelevant discourses were all but squelched; "no matter how long the Conference List," Justice Douglas recalls, "we were usually through by four-thirty or five. The discussions were short; Hughes's statements were always succinct."[11]

Chief Justice Warren did not conduct conferences in the Hughes assembly-line manner, but if Hughes was the most efficient, Warren may have been the most effective in presiding over the sessions—the "ideal" conference head, in Justice Stewart's phrase to me. He certainly allowed the Justices more leeway than Hughes had done, but that did not make him any less the leader of the conference sessions. Warren's great forte was his ability to present cases in a manner that set the tone for the discussion that followed. He would state the issues in a deceptively simple way, stripped of legal technicalities, and, where possible, relate the issues to the ultimate values that concerned him. In the face of such an approach, opposition based upon traditional legal-type arguments seemed inappropriate, almost pettifoggery. As Justice Fortas once said to me, "Opposition based on the hemstitching and embroidery of the law appeared petty in terms of Warren's basic-value approach."

As important, my own study of the conference notes in the major Warren Court decisions shows that where Warren sought to lead the Court in a particular direction, he was usually able to do so. During his tenure, the high bench was emphatically the *Warren* Court, and he, as well as the country, knew it.

Plainly, Chief Justice Warren was going to be a tough act for Warren Burger to follow. If anything, his conference performances in the *Swann* case (as well as in *Alexander v. Holmes County Board of Education,*[12] the other Burger Court case to be discussed in detail, in Chapter 4) were far from effective. Most important, he allowed the leadership in both cases to be taken from him, although he had tried to keep control of the decisions by assigning the opinions to himself.

Assigning Opinions

It has become settled by custom that it is for the Chief Justice to assign the writing of Supreme Court opinions. Aside from managing the conference, this is the most important function of the office of Chief Justice. In discharging it, a Chief Justice determines what use will be made of the Court's personnel; the particular decisions he assigns to each Justice in distributing the work load will influence both the growth of the law and his own relations with his colleagues.

The power of the Chief Justice to assign opinions probably goes back to John Marshall's day. When Marshall was appointed, the Court followed the English practice of delivery of separate opinions by each of the Justices. As soon as Marshall began to discharge his duties as Chief Justice, Beveridge's classic biography informs us,

> he quietly began to strengthen the Supreme Court. He did this by one of those acts of audacity that later marked the assumptions of power which rendered his career historic. For the first time the Chief Justice disregarded the custom of the delivery of opinions by the Justices *seriatim,* and, instead, calmly assumed the function of announcing, himself, the views of that tribunal. Thus Marshall took the first step in impressing the country with the unity of the highest court of the Nation.[13]

Marshall's innovation was strongly opposed by both Jefferson and Madison, but by 1823 even the latter recognized that it had become established procedure. "I have taken frequent occasions," Madison wrote to Jefferson in that year, "to impress the necessity of the seriatim mode;

but the contrary practice is too deeply rooted to be changed without the injunction of a law, or some very cogent manifestation of the public discontent."[14]

Since that time, it has been the uniform practice to have the decision in each case explained by an opinion of the Court delivered by one of the Justices. By the end of Marshall's tenure, too, it had become established that the Chief Justice was the one to assign opinions. In a speech on the practice under Chief Justice Taney, Marshall's successor, Justice Campbell told how the decision would be made in accordance with the Justices' votes at the conference, which also "signified the matter of the opinion to be given. The Chief Justice designated the judge to prepare it."[15]

What happened, however, when the Chief Justice was not a member of the majority? During Marshall's early years, it is probable that Marshall delivered the opinion of the Court even in cases where his personal views differed from those of the Court. Apparently the practice then was to reserve delivery of the opinion of the Court to the Chief Justice or the senior associate judge present on the bench and participating in the opinion.[16] But, as time went on, other Justices also began to deliver opinions. By Taney's day, as seen, it was settled that the Chief Justice would choose the judge to write each opinion.

In the early years of opinion assignment by the Chief Justice, he may well have assigned all opinions, even in cases where he was in dissent. It was not very long, however, before the Chief Justice's assigning power was limited to cases where he had voted with the majority. It is probable that this practice also developed under Chief Justice Taney. In his authoritative history of the Supreme Court under Taney's successor, Chief Justice Chase, Charles Fairman describes the procedure at the beginning of Chase's tenure: "The writing of opinions was assigned by the Chief Justice—save that if he were dissenting, the Senior Justice in the majority would select the one to write."[17]

Whether or not this procedure was as firmly established as this statement indicates,[18] there is no doubt that, by the present century, the uniform practice limited the Chief Justice's assigning power to cases where he was not in dissent. The situation was summarized in 1928 by former Associate Justice Hughes: "After a decision has been reached, the Chief Justice assigns the case for opinion to one of the members of the Court, that is, of course, to one of the majority if there is a division and the Chief Justice is a member of the majority. If he is in a minority, the senior Associate Justice in the majority assigns the case for opinion."[19]

This history of the Chief Justice's power to assign opinions is pertinent to our discussion of the Burger Court and its decision process in the

Swann case because, as we shall see, Chief Justice Burger did not follow the established practice in his assignment of the *Swann* opinion to himself. *Swann* was not the only case in which Burger assigned a case even though he had not voted with the majority. "He does it all the time," one of the Justices told me about the Chief Justice's assignment practice. That is doubtless an exaggeration, but Burger has assigned opinions in other cases where he was not in the majority.[20] An outstanding example was the 1973 abortion case *Roe* v. *Wade*,[21] where the Chief Justice assigned the opinion to Justice Blackmun, though it was not clear that Burger was then part of the majority.

Justice Douglas prepared a strong unpublished dissent which asserted that he, as senior member of the majority, should have assigned the abortion case. "When, however," Douglas asserted, "the minority seeks to control the assignment, there is a destructive force at work in the Court. When a Chief Justice tries to bend the Court to his will by manipulating assignments, the integrity of the institution is imperilled."[22]

Not long before, Douglas had written a memorandum to the Chief Justice complaining about the Burger assignment practice in *Swann* and another case.[23] "If the conference," Douglas wrote, "wants to authorize you to assign all opinions, that will be a new procedure. Though opposed to it, I will acquiesce. But unless we make a frank reversal in our policy, any group in the majority should and must make the assignment."[24]

The New Chief Justice

Warren E. Burger was essentially a self-made man. Growing up in St. Paul, he attended the University of Minnesota for two years and then went to night law school. To support himself he worked during the day selling life insurance. Graduated during the depression, he went into practice in St. Paul, where he also became involved in Republican politics. He was active in Harold E. Stassen's successful campaign for governor and in his later unsuccessful presidential bids. In 1952, Burger was Stassen's floor manager at the Republican convention, when Minnesota's switch to Eisenhower supplied the necessary votes for the latter's nomination.

After Eisenhower's election, Burger was appointed assistant attorney general in charge of the Claims Division of the Department of Justice. His tenure in that position is best remembered for his argument before the Warren Court defending the dismissal of a government employee on loyalty grounds, after Solicitor General Simon E. Sobeloff had refused to argue the case.[25] More important in its impact on the law would be the years Burger spent reviewing government contract debarments, which barred contractors from doing business with the government. When, in 1964, then–Circuit Judge Burger wrote a precedent-shattering opinion invalidating a contract debarment without notice and hearing,[26] he did so with the confidence of one who knew a great deal about this subject. All too frequently, he recalls, he would see debarment orders made on their face by the secretary of the navy, but perusal of the files would show that the real decision had been made by some lieutenant j.g. far down the line.

In 1956 Burger was appointed to the U.S. court of appeals in Washington, D.C., where he developed a reputation as a conservative, particularly in criminal cases. Then, in 1969, came what Justice Harlan was to term his "ascendancy to the Jupitership of Mount Olympus."[27] President Nixon selected Burger as Chief Justice Warren's successor. In a session attended by the president, Burger was sworn in by the retiring Chief Justice on June 23, 1969, Warren's last day on the bench.

In appearance, Burger is the casting director's ideal of a Chief Justice. On the bench and at public functions, Burger's white-maned, broad-shouldered presence is an almost too perfect symbol of the law's dignity. His critics contend that he stands too much on the dignity of his office and is, more often than not, aloof and unfeeling. Intimates, however, stress his courtesy and kindness and assert that it is the office, not the man, that may make for a different impression. They speak of his wry humor, which is rarely displayed in public. After Justice Blackmun, who grew up with the Chief Justice, was appointed, Burger sent Justice Harlan a newspaper clipping that read: "The new Chief Justice Warren Burger has his fellow Minnesotan, thus making it finally a dull Court." Under this, Burger had written, "Well, that's the way it is with people from Minnesota!"[28]

From his "Middle Temple" Cheddar (made, according to the Burger recipe, "with English Cheddar, Ruby Port, Brandy and a few other ingredients")[29] to the finest clarets, the Chief Justice is somewhat of an epicurean. One of the social high points of a 1969 British-American conference at Ditchley, Oxfordshire, attended by the newly appointed Chief Justice, was the learned discussion about vintage Bordeaux that took place between Burger and Sir George Coldstream, head of the Lord Chancellor's office, who was in charge of the wine cellar at his Inn of Court. The Chief Justice was particularly proud of his coup in snaring some cases of rare Lafite in an obscure Washington wine shop.

As Chief Justice, Burger has been more effective as a court administrator and as a representative of the federal courts before Congress than as a molder of Supreme Court jurisprudence. Looking back at the Warren years, Justice Byron White pointed out that, as far as relations with Congress are concerned, "Things have changed . . . for the better as far as I can see" under Burger. As White recalls it, "Chief Justice Warren did have such a problem with the civil rights thing, and with prayers and reapportionment. Congress was in such a terrible stew that his name was mud [there], which rubbed off on all of us." Under Burger the situation has been different. Few Chief Justices have had better relations with Capitol Hill.

As far as court administration has been concerned, Burger has played a more active role than any Court head since Chief Justice Taft. His administrative efforts have ranged from efforts at fundamental changes, such as his active support of the creation of a new court of appeals to screen cases that the Supreme Court would consider, to attention to what some might consider petty details in the Court's functioning.

Before Burger's tenure, the Court did not have any photocopy machines. Documents had to be typed with as many carbon copies as needed. Thus, the memoranda on *in forma pauperis* (or Miscellaneous Docket) petitions prepared by Chief Justice Warren's law clerks were typed in originals and carbons were sent to each Justice. They were called "flimsies," because making eleven copies required the use of very thin paper; the junior Justices, who received the last copies, often had difficulty reading them.

This was changed when Chief Justice Burger sent around an August 7, 1969, memorandum: "The necessary steps are now being taken toward acquiring a Xerox machine in the building." The new copiers "will be utilized primarily in preparation and distribution of the Miscellaneous Docket memoranda ('flimsies' as they are commonly called.)"[30] Since that time, ample copiers have been provided for the Justices and Court staff and word processors and a computer have more recently been introduced.[31]

Comparable innovations made by the new Chief Justice have concerned new lighting[32] (including fluorescent lights for the conference room) and a new winged bench to replace the more traditional straight bench that had always been used in the courtroom. In a February 4, 1971, *Memorandum to the Conference*, Burger wrote that everyone was in favor of the "change with the possible exception of Justice Black whose position can probably be described as 'take it or leave it.' "[33]

Like his predecessor, Chief Justice Burger has been protective of the dignity of his office. When he moved into his chambers, he was dismayed to find that his office was smaller than the one he had had at the court of appeals. Next door was the elegant conference room, which could serve admirably as a ceremonial office for the Court head. Burger did not, however, go so far as to take over the conference room. He placed an old desk in the room and moved the conference table to one side. This may have irritated the Justices,[34] but no formal objection was made and the conference room now serves as the Chief Justice's receiving room; visitors, often there for afternoon tea, now enjoy the splendid oak-paneled chamber.

At the time the *Swann* case was appealed to the Supreme Court, the new Chief Justice was completing his first year. Soon thereafter, on July 15, 1970, he sent a long letter to Justice Harlan.[35] Headed "Personal & Private" in Burger's writing, the letter summed up his impressions "in the first year of operations." The "things which disturb me . . . ," Burger wrote, "are largely some of the 'chronic' problems of the Court, particularly the 'supermarket' or 'production line' aspect the Court has drifted into over the years. . . . With the same manpower of Justices the Court is now trying to do 400%, more or less, of what was done in Hughes' day. It can't be done." Burger conceded that his emphasis on this point might be "dogmatic"—"but dogmatic or not, I intend to dedicate my remaining years to changing this horrendous pattern which has gravely affected the work of the Court."

As the Chief Justice saw it, "We are like the farmer who decided to lift his new colt every morning and evening on the theory he could always muster strength to carry 'a few more pounds.' The 'no lift' point is not in the future; perhaps it is about 10–15 years past, and we just *think* we are doing the job properly." His first year, Burger's letter concluded, "has confirmed the worst fears I had as to why much of the Court's work is not up to what it ought to be if we are to preserve the great traditions. Sad but true, and I am much too old to pretend to be starry-eyed about it. . . . If this sounds pessimistic, be sure also that my pessimism does not alter my determination to change the picture I see."

Black's "Apostasy"

When Chief Justice Burger was appointed, the senior Justice was Justice Black, who had been on the Court since 1937. Hugo L. Black was one of the most colorful Justices in our history—as strong an individualist as any to serve on the Court. Black would always insist on doing things his own way. The ritual intonation that opens each Supreme Court session ends with, "God save the United States and this Honorable Court." One year a new member of the Court began bowing his head when those words were reached. The Justice beside him soon followed suit. It spread down the line until everyone bowed his head, except Justice Black, who continued to look straight out at the audience.[36]

By the time Burger became Chief Justice, the furor that had surrounded Black's appointment because of the disclosure that he had once been a member of the Ku Klux Klan seemed an echo from another day. "At every session of the Court," a *New York Times* editorial thundered after Black's Klan membership had been revealed, "the presence on the bench of a justice who has worn the white robe of the Ku Klux Klan will stand as a living symbol of the fact that here the cause of liberalism was unwittingly betrayed."[37] Ten years later, Black himself was the recognized leader of the Court's liberal wing.

Black never forgot his origins in a backward Alabama rural county. Half a century later, he described a new law clerk from Harvard as "tops in his class though he came from a God-forsaken place—worse than Clay County."[38] But his Alabama drawl and his gentle manners masked an inner firmness found in few men. "Many who know him," wrote Anthony Lewis when Black turned seventy-five, "would agree with the one-time law clerk who called him 'the most powerful man I have ever met.' "[39] Though of middling height and slight build, Black always amazed people with his physical vitality. He is quoted in *The Dictionary of Biographical Quotation* as saying, "When I was forty my doctor advised me that a man in his forties shouldn't play tennis. I heeded his advice carefully and could hardly wait until I reached fifty to start again."[40]

Black's competitive devotion to tennis became legend. Until he was eighty-three, he continued to play several sets every day on the private court of his landmark federal house in the Old Town section of the Washington suburb of Alexandria. He brought the same competitive intensity to his judicial work. According to his closest colleague, Justice Douglas, "Hugo Black was fiercely intent on every point of law he presented."[41] Black was as much a compulsive winner in the courtroom as on the tennis court. "You can't just disagree with him," acidly commented his great Court rival, Justice Jackson, to *New York Times* columnist Arthur Krock. "You must go to war with him if you disagree."[42] Black would fight bitterly on the issues that concerned him, such as the First Amendment. His combative approach is well reflected in the stands he took and the strong draft dissents he wrote in *Alexander v. Holmes County Board of Education*[43] as well as the *Swann* case itself.

Yet, if impact on the law is a hallmark of the outstanding judge, few occupants of the bench have been more outstanding than Black. It was Justice Black who fought for years to have the Court tilt the Constitution in favor of individual rights and liberties and who was, next to Chief Justice Warren himself, the leader in what a Justice once termed "the most profound and pervasive revolution ever achieved by substantially peaceful means."[44] Even where Black's views have not been adopted literally, they have tended to prevail in a more general, modified form. Nor has his impact been limited to the Black positions that the Court has accepted. It is found in the totality of today's judicial awareness of the Bill of Rights and the law's new-found sensitivity to liberty and equality.

More than anything else, Justice Black brought to the highest Court a moral fervor rarely seen on the bench. A famous passage by Justice Holmes has it that the black-letter judge will be replaced by the man of statistics and the master of economics.[45] Black was emphatically a judge who still followed the black-letter approach in dealing with the constitutional text. "That Constitution," he said, "is my legal bible. . . . I cherish every word of it from the first to the last."[46] The eminent jurist with his dog-eared copy of the Constitution in his right coat pocket became a part of the contemporary folklore. In protecting the sanctity of the organic word, Justice Black displayed all the passion of the Old Testament prophet in the face of the graven idols. His ardor may have detracted from the image of the "judicial." But if the Justice did not bring to constitutional issues that "cold neutrality" of which Edmund Burke speaks,[47] zeal may have been precisely what was needed in his sometimes lonely quest to place a literal reading of the Bill of Rights at the vital center of our constitutional law.

The Black who sat on the Burger Court, however, was no longer the leading liberal on the Court. In his eighty-third year and thirty-second term on the Court, his health had begun to give way. He had recently suffered a stroke while playing tennis and had finally given up the game. But the Alabaman's fundamentalist approach to the Constitution did not permit him to adopt the expansive approach toward individual rights now followed by some of the Justices.

Black stood his constitutional ground where the rights asserted rested on specific provisions, such as the First Amendment or the Fifth Amendment privilege against self-incrimination, but when he could not find an express constitutional base, Black was unwilling to create one to meet a new need. This limited approach would lead Black to his hostility toward school busing in the *Swann* case. "Where does the word *busing* appear in the Constitution?" Black is said to have asked his law clerks.[48]

Black's "apostasy" had become apparent during the last years of the Warren Court, when he had refused to join the Court in recognizing the new right of privacy, because he could not find an express constitutional foundation for it,[49] and had voted to uphold convictions of certain civil rights demonstrators, giving greater weight to the constitutional guarantees protecting property rights, including the owner's right to limit access to his property.[50] During his early years on the Court, Black was perceived to be far out on the liberal wing. First the Court moved toward him, and then past him, leaving the old liberal closer to the conservative wing.

Chief Justice Warren had been well aware of the change in Black's position vis-à-vis the rest of the Court and was more tolerant of it than some of his colleagues. When others complained to the Chief Justice about Black's senility (Black turned eighty during the 1965 Term), Warren would chuckle and say, "No, Hugo just wants to be buried in Alabama."

The Liberal Core

The Court that would decide the *Swann* case contained a liberal core composed of Justices Douglas, Brennan, and Marshall. The first two played a prominent part in the *Swann* decision process and deserve more than cursory mention in this survey of the Justices.

William O. Douglas was an imposing physical presence on the Court. A sinewy six-footer, he seemed the personification of the last frontier, the down-to-earth Westerner whose granite-hewed physique always seemed out of place in Parnassus. More than that, Douglas was the Court's Horatio Alger, whose early life was a struggle against polio and poverty. Told that he would never walk, he became a noted sportsman. Riding east on a freight car to enroll at Columbia Law School, with six cents in his pocket, he became an eminent law professor, chairman of the Securities and Exchange Commission, and, at the age of forty, a Supreme Court Justice.

But the real Douglas was different from his public image. Douglas was the quintessential loner, a lover of humanity who did not like people. No one on the Court, neither among the Brethren nor the law clerks, was really close to the strapping Westerner. The clerks particularly describe Douglas as the coldest of the Justices. It was an event when Douglas stopped to say hello in the Court corridors. His severity toward his own clerks became legend. At times, his behavior toward clerks was downright cruel. Once, when a major university wanted to invite Douglas to speak, it asked one of Chief Justice Warren's clerks, a recent graduate, to make the necessary approach. The clerk arranged an appointment and proudly introduced the president of the student body, who started to tell the Justice what they wanted. At this, Douglas broke in and snapped, "Why are you bothering me with that? Talk to my agent; the fee is $3,000, and he'll handle it."

On the bench, as in his personal life, Douglas was a maverick, going his own way regardless of the feelings of the Brethren. "Bill's headstrongness sometimes cannot be coped with," wrote Douglas's second wife to Black one summer.[51] Douglas would stick to his own views, quick to state his own way of deciding in concurrence or dissent. It made little difference whether he carried a majority or stood alone. He would rarely stoop to lobbying for his position and seemed more interested in making his own stand public than in working to get it accepted. As one law clerk put it to me, "Douglas was just as happy signing a one-man dissent as picking up four more votes."

There is no doubt that Douglas had a brilliant mind, but he was erratic. He could whip up opinions faster than any of the Justices and did almost all the work himself, relying less on his clerks than anyone else. But his opinions were too often unpolished, as though he lacked the patience for the sustained work involved in transforming first drafts into finished products. Less than total interest in the Court's work was, indeed, apparent in Douglas's whole attitude. The peripatetic Justice spent much time writing nonlegal books and doing things completely unrelated to the Court, mostly traveling and mountain climbing.

Until Chief Justice Warren's last years, Douglas had been as close to Justice Black as to anyone on the Court. "If any student of the modern Supreme Court took an association test," wrote Black's son in his book about his father, "the word 'Black' would probably evoke the response 'Douglas' and vice versa."[52] Black himself recognized this. Declining a 1958 invitation to write an article about Douglas, Black wrote, "You perhaps know without my stating it that I have the very highest regard for Justice Douglas as a friend and as a member of this Court. In fact, our views are so nearly the same that it would be almost like self praise for me to write what I feel about his judicial career."[53]

All this had changed with Black's already discussed "apostasy." Now Douglas was the senior member of the Court's liberal bloc; but he was scarcely the man to lead that bloc in the Burger Court. That role was assumed by Justice Brennan, who had become the Chief Justice's principal lieutenant during the Warren years. Now, under the new Chief Justice, it was Brennan who was to lead the Justices in trying to preserve the heritage of the Warren Court.

Before his 1956 appointment by President Eisenhower, William J. Brennan, Jr., had been a judge in New Jersey for seven years, rising from the state trial court to its highest bench. He was the only Justice to have served as a state judge.

"One of the things," Justice Frankfurter once said, "that laymen, even lawyers, do not always understand is indicated by the question you hear so often: 'Does a man become any different when he puts on a gown?' I say, 'If he is any good, he does.' "[54] Certainly Justice Brennan on the supreme bench proved a complete surprise to those who saw him as a middle-of-the-road moderate. He quickly became a firm adherent of the activist philosophy and a principal architect of the Warren Court's jurisprudence. Brennan had been Frankfurter's student at Harvard Law School. Yet, if Frankfurter expected the new Justice to continue his pupilage, he was soon disillusioned. After Brennan had joined the Warren Court's activist wing, Frankfurter supposedly quipped, "I always encouraged my students to think for themselves, but Brennan goes too far!"

Brennan soon became Chief Justice Warren's closest colleague. The two were completely dissimilar in appearance: The Justice was small and feisty, almost leprechaun-like in appearance, yet he had a hearty bluffness and an ability to put people at ease. Brennan's unassuming appearance and manner mask a keen intelligence. He is perhaps the hardest worker on the Court. Unlike Justices like Douglas, Brennan has always been willing to mold his language to meet the objections of some of his colleagues, a talent that would become his hallmark on the Court and one on which Chief Justice Warren would rely frequently. It was Brennan to whom the Chief Justice assigned the opinion in some of the most important cases decided by the Warren Court.

Now, in the Burger Court, Brennan was no longer the Chief Justice's trusted insider. In the *Swann* case, indeed, it was the new Chief Justice who was to be a principal opponent of the Brennan view, a view that held *Swann* to be merely another occasion for the implementation of Warren Court desegregation decisions.

Conservative and Moderate Justices

During the early years of the Warren Court, the opposition to the activist view increasingly assumed by the Court was led by Justice Frankfurter. With Frankfurter's 1962 retirement, the principal voice calling for judicial restraint was that of Justice John Marshall Harlan. He was the grandson of the Justice John Marshall Harlan who had written the famous dissent in the 1896 case of *Plessy* v. *Ferguson*.[55] As soon as he took his place on the bench, the second Justice Harlan made Court history as the only descendant of a Justice to become a Justice. The first Harlan had been a judicial maverick and an outspoken dissenter. The second Harlan took a more cautious approach to the judicial function, which reflected his background. Educated at Princeton, then a Rhodes scholar, Harlan spent many years as a successful attorney with a leading Wall Street law firm. In 1954 he was appointed to the U.S. Court of Appeals for the Second Circuit, on which he served briefly before being named to the Supreme Court.

Harlan looked like a Supreme Court Justice. Tall and erect, with sparse white hair, conservatively dressed in his London-tailored suits, with his grandfather's gold watch chain across the vest beneath his robe, he exuded the dignity associated with high judicial office. Yet underneath this dignified surface was a warm nature that enabled him to be close friends with those with whom he disagreed intellectually, notably Justice Black. Visitors could often see the two Justices waiting patiently in line in the Court cafeteria. The two were a study in contrasts: the ramrod-straight patrician with a commanding presence and his slight, almost wispy colleague who always looked like the lively old southern farmer.

The term most frequently used to describe Harlan by those who knew him is "gentleman." To the Justices, Harlan always appeared the quintessential patrician, with his privileged upbringing and Wall Street background. "You are expert on all things English," reads a handwritten note from Burger to Harlan on a London speech the Chief Justice was to deliver. "Do you have any comments?"[56]

Harlan was one of the best, if not the best, lawyer on the Court and the one most interested in the technical aspects of the Court's work. He became a sound, rather than brilliant, Justice who could be relied on for learned opinions that thoroughly covered the subjects dealt with, though they degenerated at times into overlong law review articles of the type Justices all too often write.

When Chief Justice Burger took his Court seat, he had reason to expect support from Harlan. The Justice shared Burger's concern over "the 'horse and buggy' conditions under which the federal judiciary, and particularly this Court, are now operating"[57] and he had been the leading conservative in the Warren Court. There was, however, one area of the law where Harlan was anything but conservative. If there was any one person whom Harlan revered, it was his grandfather, the Justice who is best remembered for his ringing condemnation of racial segregation in *Plessy* v. *Ferguson*. There is a revealing correspondence between Justices Frankfurter and Harlan in which Frankfurter denigrates Harlan I and his *Plessy* dissent.[58] At this, Harlan II bristled and wrote, "I think you push things too far."[59] In cases involving racial equality, Harlan considered himself his grandfather's direct heir. In a case like *Swann*, he would be as vigorous in supporting desegregation as Justices Brennan and Douglas, and would join in their opposition to the Burger efforts to weaken Judge McMillan's school busing order.

Except for Justice Douglas, Justice Potter Stewart was the youngest Court appointee in over a century (he was 43 when selected; Douglas had been 40). Stewart's youth and handsome appearance added an unusual touch to the highest bench, showing that it need not always be composed of nine old men. Stewart also came from a distinguished legal family. He had attended Hotchkiss and Yale, taken a postgraduate year at Cambridge, England, and been an honors graduate of Yale Law School. He had been active in municipal politics in Cincinnati, with two terms on the city council. President Eisenhower had appointed him to the federal court of appeals in 1954, four years before elevating him to the Supreme Court.

In his early years on the Court, Stewart tended to be the "swing man" between the Warren and Frankfurter blocs. During the 1958 and 1959 Terms, the two blocs were evenly divided and Stewart cast the key vote in many cases, voting now with the one and now with the other group.[60] With Frankfurter's 1962 retirement, Warren and his supporters gained the upper hand. Stewart remained in the center as the Court moved increasingly to the left. At the end of Warren's tenure, he continued as the Court's leading moderate—though, according to Justice Douglas, by that time "Stewart and Harlan were the nucleus of the new conservatism on the Court."[61]

Anyone who meets Stewart is surprised by his vigor and clearly expressed views. They contrast sharply with his public image as an indecisive centrist without clearly defined conceptions. Unlike Justices Black and Frankfurter, Stewart never acted on the basis of a deep-seated philosophy regarding the proper relationship between the state and its citizens. When asked if he was a "liberal" like Black or a "conservative" like Frankfurter, he answered, "I am a lawyer," and went on to say, "I have some difficulty understanding what those terms mean even in the field of political life. . . . And I find it impossible to know what they mean when they are carried over to judicial work."[62]

At times, Stewart's aptness for the pungent phrase, apparent to anyone who engaged him in private conversation, broke through the robe's restraint. Most people have heard of his famous statement, in a 1964 case about hard-core pornography, when he maintained that, though he would not attempt to define it, "I know it when I see it."[63] Stewart would often make such pointed remarks, though infrequently on the bench. Another example occurred during the argument of an antitrust case involving the Brown Shoe Company. Brown's attorney argued that all men's shoes did not compete for sales with all other men's shoes; thus, dress shoes did not compete for sales with casual shoes. The solicitor general responded that they did, pointing to the example of Brown's president, who had come to court to testify wearing dress shoes one day and casual shoes the next. Stewart grinned and declared, "Maybe it was direct examination one day and cross-examination the next."[64]

The Other Justices

The remaining three Justices did not perform as important a role in the *Swann* decision process and can be dealt with more briefly. One (Justice Byron R. White) is perhaps best classified as a conservative, one (Justice Thurgood Marshall) as a liberal, and one (Justice Harry A. Blackmun) as an unknown quantity, but at that time most likely to be a conservative in the Burger mold.

When he was selected by President Kennedy in 1962, Byron R. White was certainly not the typical Supreme Court appointee. He was known to most Americans as "Whizzer" White, the all-American back who became the National Football League rookie of the year in 1938. On graduation from law school after the war, White clerked for Chief Justice Vinson (he was the first former law clerk to become a Justice himself). After practicing law in his native Colorado for fifteen years, he became chairman of the nationwide Citizens for Kennedy in 1960 and moved to Washington as deputy attorney general, the position he held at the time of his Court appointment.

Physically, Justice White is most impressive. At six feet two and a muscular 190 pounds, he has maintained the constitution that made him a star quarterback. Even as a Justice, White has retained his athletic competitiveness, never hesitating to take part in the clerks' basketball games in the gymnasium at the top of the Court building.

On the bench, White, like Stewart, has defied classification. He, too, tends to take a lawyerlike approach to individual cases, without trying to fit them into any overall judicial philosophy. During the Warren years, White was never close to the Chief Justice. "I wasn't exactly in his circle," White told me. Certainly White never became a member of the Warren Court's inner circle. He went his own way, voting against the Chief Justice as often as not. Toward the end of Warren's tenure, he tended to vote most frequently with Justices Harlan and Stewart and was considered one of the more conservative Justices.

Thurgood Marshall, on the other hand, was almost always to be found in the ranks of the Warren supporters. Marshall himself gave a racial dimension to the Horatio Alger legend. Great-grandson of a slave and son of a Pullman car steward, Marshall was the first black appointed to the highest Court. He had headed the NAACP Legal Defense Fund's staff for over twenty years and had been chief counsel in the *Brown* school segregation case.

Marshall had been appointed to the U.S. Court of Appeals for the Second Circuit in 1961 and had been named solicitor general in 1965. When Marshall balked at the latter appointment, it is said that President

Johnson told him, "I want folks to walk down the hall at the Justice Department and look in the door and see a nigger sitting there."[65] The president elevated Marshall to the Supreme Court for the same reason. As Justice Douglas put it in his *Autobiography,* "Marshall was named simply because he was black."[66]

After Marshall had served on the U.S. court of appeals in New York for a year, Henry J. Friendly, the outstanding judge on that court wrote to Justice Frankfurter in January 1962: "TM seems easily led. I do not have the feeling that he realizes the difficulties of his job and is burning the midnight oil in an effort to conquer them. . . . All this makes life fairly easy for him, save when he is confronted with a difference of opinion, and then he tosses a coin."[67] A month later Friendly wrote again: "I continue to be alarmed by Marshall's willingness to arrive at quick decisions on issues he does not understand."[68] The Friendly estimate was equally valid after Marshall's elevation to the Supreme Court. On the high bench, he has been a firm vote for the liberal bloc. In the Burger Court he has served as a judicial adjunct to Justice Brennan. The law clerks, it is said, took to calling Marshall "Mr. Justice Brennan-Marshall."[69]

The remaining Justice, Harry A. Blackmun, was the second appointed by President Nixon. He had served eleven years on the U.S. Court of Appeals for the Eighth Circuit when the call came from Washington in 1970. Though born in Illinois, he was raised in Minnesota. He went to grade school with Warren Burger and the two remained close friends thereafter, with Blackmun serving as best man at Burger's wedding. After graduation from Harvard Law School, Blackmun served as a law clerk in the court of appeals to which he would later be appointed, spent sixteen years with a Minneapolis law firm, and was counsel to the famous Mayo Clinic for almost ten years.

When the *Swann* case came to the Supreme Court, Blackmun was as yet untested as a Justice. He was still largely under the Chief Justice's influence. During his first year, Blackmun and Burger were almost always on the same side. The press had come to call them the "Minnesota Twins," after the baseball team from their hometowns (the Twin Cities of Minneapolis and St. Paul). The attitude toward the two within the Court is indicated by some anti-Burger doggerel circulated to some of the Justices in 1971:

. . . Burger is being aided and abetted
By his yes-man, Blackmum [sic], for whom he stood up when wedded,
Or was it the other way around, Blackmum [sic] standing up for Berger [sic]?
Regardless, it is plain that they still honor their old time merger:

YOU STAND UP FOR ME AND I'LL STAND UP FOR YOU AND TOGETHER
WE WILL SOW DISORDER
TO HELL WITH THE CONSTITUTION, WE ARE FOR NIXON'S LAW AND
ORDER.[70]

How the Court Operates

To understand how the Supreme Court dealt with the *Swann* case, one must understand how the Court itself operates. The Justices sit from October to late June or early July, in annual sessions called terms, with each term designated by the year in which it begins. Cases come to the Court from the U.S. courts of appeals and the highest state courts, either by appeals or petitions for writs of certiorari. Technical rules govern whether an appeal or certiorari must be sought. The Justices have virtually unlimited discretion in deciding whether to take an appeal or grant certiorari (or cert, as it is usually called in the Court). Each year the Justices decide to hear only a fraction of the cases presented to them. Thus, in the 1969 Term, when the *Swann* plaintiffs filed their certiorari petition asking the Supreme Court to overturn the court of appeals decision, the Court agreed to hear only 150 cases out of 4,202 presented. In 1955, Justice Frankfurter had asked another Justice, "Wouldn't you gladly settle for one in ten—such is my proportion—in granting petitions for certiorari?"[71] By the time Chief Justice Burger was appointed, the portion granted had declined to one in twenty-eight.

Following an unwritten rule, when at least four of the nine justices vote to take a case, certiorari is granted or the appeal is taken. If, however, the case elicits fewer than the required four votes, the case in question is over and the last decision of the state court or lower federal court becomes final. In recent years some Justices have urged that changes be made in the Rule of Four, on the ground that it results in the Court agreeing to hear too many cases. In his already quoted letter to Justice Harlan, Chief Justice Burger wrote, "The hurried, casual granting of certs in worthless cases appalls me. I hope you will support a proposal I shall again make in October that when cert is granted by less than six votes, the 'granters' will take a second look and memo the Conference on why the case is certworthy. That would have saved us from 8 to 16 cases this past year."[72]

More recently, Justice Stevens has proposed that five votes be required to grant certiorari.[73] Interestingly, if a five-vote rule had been in effect during the Warren Court years, at least one of the most important Warren Court decisions, *Baker* v. *Carr*,[74] which Chief Justice Warren once described as "the most important case of my tenure on the Court,"[75] would never have been decided. Only four Justices voted to hear that case.

For those few cases the Supreme Court agrees to take, written briefs will be submitted by the opposing lawyers, and then the attorneys for both sides will appear for oral argument. The arguments are presented publicly in the ornate courtroom. Each side usually has half an hour, and the time limit is strictly observed. Once a lawyer was arguing his case in the last half hour before noon. He was reading to the Justices from his notes and did not notice the red light go on at his lectern, signaling that his time had expired. Finally, he looked up. The bench was empty; the Justices had quietly risen, gathered their black robes, and gone to lunch.

As far as the public is concerned, the postargument decision process in the Court is completely closed. The next time the outside world hears about the case is when the Court is ready to publicly announce its decision; simultaneously, the majority opinion and any dissents or concurrences are distributed. But in that interim period between oral argument and the announcement of the Court's decision, much has gone on. First, the Justices have "conferenced." When Burger became Chief Justice, these conferences were held on Fridays. More recently, Wednesday sessions have been held as well. The privacy of the conference is one of the most cherished traditions at the Court. Only the nine Justices may attend. In addition to the conference discussion, ideas are exchanged by the Justices through the circulation of draft opinions and memoranda. Such memos, sent to all the Justices, are usually titled *Memorandum to the Conference.*

After the vote is taken at the conference, the case is assigned by the Chief Justice, if he is in the majority, either to himself or to one of the Justices for the writing of an opinion of the Court. As already stressed, if the Chief Justice is not in the majority, the senior majority Justice assigns the opinion. Justices who disagree with the majority decision are free to write or join dissenting opinions. If they agree with the result but differ on the reasoning, they can submit concurring opinions. Opinions are usually issued in the name of individual Justices. Sometimes per curiam (literally, "by the court") opinions are issued in the name of the Court as a whole. That, we shall see in Chapter 4, is what happened in the *Alexander* case,[76] the first school desegregation case decided by the Burger Court.

The last stage is the public announcement of decisions and the opinions filed by the Justices. The custom used to be to have decisions announced on Mondays (a tradition that began in 1857); hence, the press characterization of "decision Mondays." In 1965, this was changed to announcing decisions when they were ready.

When decisions are announced, the Justices normally read only a summary of their opinions, especially when they are long. But some insist on reading every word, no matter how much time it takes. On June 17, 1963, Justice Clark was droning through his lengthy Court opinion in the case involving the constitutionality of Bible reading in public schools. Justice Douglas, who could stand it no longer, passed Justice Black a plaintive note: "Is he going to read all of it? He told me he was only going to say a few words—he is on p. 20 now—58 more to go—Perhaps we need an anti-fillibuster rule as badly as some say the Senate does."[77]

3

The Warren Court Decisions

An attempt to draft fundamental law on a tabula rasa may turn out favorably in Greek myth, but in our society the judge must start with existing law as it has evolved through prior cases. New law can be understood only in light of its legal setting, particularly the precedents that have led up to it.

In ruling with the plaintiffs in *Swann* Judge McMillan made a far-reaching school desegregation decision, including an order for extensive busing. As he saw it, however, the decision was compelled both by the facts in this particular case and by Supreme Court precedents. The precedents he saw as determinative were those decisions of the Warren Court treating the relationship between segregation and educational opportunity, starting with the landmark decision, *Brown* v. *Board of Education*.[1]

Brown I

If, as Disraeli tells us, a precedent embalms a principle, the key principle in school segregation cases was established in 1954, in the *Brown* case, the watershed constitutional case of the present century. *Brown* was, of course, the case that struck down school segregation and destroyed the legal foundation that supported the Jim Crow system that had become the dominant feature in southern life. Before *Brown*, the leading case had been *Plessy* v. *Ferguson*,[2] where the Supreme Court had upheld a state law requiring "equal but separate accommodations for the white and colored races." The subsequent structure of racial discrimination in much of the country was built upon the "separate but equal" doctrine approved by the Court.

The *Brown* decision completely changed the legal picture. The opinion of Chief Justice Warren categorically repudiated the notion that school segregation could ever be consistent with equal protection. *Plessy v. Ferguson* was flatly overruled. That case had required Court denial of "the assumption that the enforced separation of the two races stamps the colored race with a badge of inferiority." "If this be so," the Court said, "it is not by reason of anything found in the act, but solely because the colored race chooses to put that construction upon it."[3] From his first connection with the *Brown* case, Chief Justice Warren rejected this *Plessy* reasoning. In his initial conference presentation on December 12, 1953, Warren indicated that he saw the question before the Court in terms of the moral issue. *Plessy*, he told the conference, cannot "be sustained on any other theory [than] a concept of the inherent inferiority of the colored race."[4] In his *Brown* opinion, the Chief Justice went further and declared that segregation inevitably contributes to the inferiority of educational opportunity for blacks. "Whatever may have been the extent of psychological knowledge at the time of *Plessy v. Ferguson,* this finding is amply supported by modern authority. Any language in *Plessy v. Ferguson* contrary to this finding is rejected."[5] This statement was supported by footnote 11 of the opinion, which listed seven works by social scientists and soon became the most controversial note in a Supreme Court opinion.[6]

Despite the commentary and controversy caused by footnote 11, the Warren draft opinions show that Warren did not base his decision on the social science studies cited in the footnote. As I have shown elsewhere,[7] footnote 11 was inserted into the opinion at a later stage by one of Warren's law clerks, and the Justices did not rely on the sociological studies cited in the footnote in coming to their unanimous decision. To the Chief Justice at least, the *Brown* decision was essentially a moral judgment compelled by the ultimate human values involved.[8] As he had declared to the conference, "I don't see how in this day and age we can set any group apart from the rest and say that they are not entitled to exactly the same treatment as all others."[9]

Such an approach placed the supporters of *Plessy* in the position of appearing to subscribe to racist doctrine. Phrasing the issue in terms of black inferiority undercut the whole basis upon which *Plessy* rested. It also rendered irrelevant the contemporary criticisms of Warren's opinion for its alleged lack of legal craftsmanship. *Brown* was so clearly right in its conclusion that segregation denies educational equality that one wonders whether a more elaborate demonstration was really necessary. After all, to anyone at all familiar with the techniques of racial discrimination, the device of state-forced separation—whether by confinement of Jews to the ghetto, exclusion of untouchables from the temple, or segregation of blacks—can mean only one thing, inequality under the law. The problem was created by the fact that the Court was not faced simply with disposing of a law that violated the Equal Protection Clause of the Fourteenth Amendment, but with a previous Supreme Court decision that had found laws passed after the amendment had been enacted not to have violated the amendment.

Brown II

The *Brown* decision of May 17, 1954, established the unqualified principle that school segregation violates the Equal Protection Clause. As far as the plaintiffs were concerned, however, the principle was without immediate practical effect, for the decision did not grant them any remedy to correct the now-established violation of their constitutional rights. As the *Atlanta Constitution* pointed out in a May 18 editorial, "The [C]ourt decision does not mean that Negro and white children will go to school this fall. The court itself provides for a 'cooling off' period. Not until next autumn will it begin to hear arguments . . . on how to implement the ruling."[10]

The Court had to determine not only what remedy to adopt to bring about the relief sought but what would be the scope of the remedy adopted by it: Should relief be granted only to the plaintiffs or should the remedy extend to all others similarly situated? The *Brown I* opinion assumed that "these are class actions" and concluded that "plaintiffs and others similarly situated" had been deprived of equal protection "by reason of the segregation complained of."[11] The draft *Brown II* opinion also applied to "the plaintiffs and those similarly situated in their respective school districts." Primarily because of the opposition of Justices Black and Douglas, the final *Brown II* opinion restricted relief to "the parties to these cases only."[12]

As far as the remedy itself was concerned, the *Brown II* Court had three main choices: (1) to issue what Justice Frankfurter called a "bare bones" decree enjoining exclusion of the plaintiffs from the white schools;[13] (2) to appoint a master to work out a detailed decree to be issued by the Court; or (3) to leave the enforcement problem to the district courts. The Court chose the third alternative, largely because of the conviction among the Justices that, as a memorandum by Justice Burton stressed, there should be "the allowance of time for . . . enforcement" and the Court should not act as a school board.[14] So far as we know, all the Justices agreed with the view stated in a Frankfurter memorandum:

"The one thing one can feel confidently is that this Court cannot do it directly."[15]

The alternative, as the Chief Justice had indicated earlier, "was to send the cases back to the trial court for the entry of appropriate decrees."[16] This would leave the enforcement to the district courts, while giving them only broad guidelines to follow in framing their decrees in individual cases. The guidance was contained in the *Brown II* opinion. Following the Chief Justice's suggestion, the Court decided to issue only an opinion listing the factors to be taken into account by district courts, rather than a formal decree.

Flexibility in enforcement was the keynote of the Warren *Brown II* opinion. The opinion stressed "the complexities arising from the transition to a system of public education freed of racial discrimination." The primary responsibility "for elucidating, assessing, and solving these problems" was in the school authorities. But the courts would have to consider whether actions of the authorities constituted good-faith implementation. The "courts which originally heard these cases can best perform this judicial appraisal,"[17] and the cases were remanded to them for enforcement.

In fashioning enforcement decrees, the *Brown II* opinion went on, the courts should be guided by equitable principles. They should balance the plaintiffs' interest in admission to the schools "as soon as practicable on a nondiscriminatory basis" with "the public interest in the elimination of such obstacles in a systematic and effective manner." However, while the courts should give due weight to public and private considerations, they "will require that the defendants make a prompt and reasonable start toward full compliance with our May 17, 1954, ruling." Once that start has been made, additional time may be necessary. In determining whether such time is needed,

> the courts may consider problems related to administration, arising from the physical condition of the school plant, the school transportation system, personnel, revision of school districts and attendance areas into compact units to achieve a system of determining admission to the public schools on a nonracial basis, and revision of local laws and regulations which may be necessary in solving the foregoing problems.[18]

Then, instead of a decree, the opinion ended by reversing the judgments denying plaintiffs relief and remanding the cases "to the District Courts to take such proceedings and enter such orders and decrees consistent with this opinion as are necessary and proper to admit to public schools on a racially nondiscriminatory basis with all deliberate speed the parties to these cases."[19]

Deliberate Speed or Indefinite Delay

When the *Brown II* opinion declared that the lower courts were to ensure that blacks be admitted to schools on a nondiscriminatory basis "with all deliberate speed," it led to learned controversy on the origins of that phrase.[20] More important, it meant that immediate relief was not to be granted for the constitutional violations that plaintiffs had suffered. *Brown I* had ruled that the plaintiffs had an unequivocal constitutional right to attend nonsegregated schools. Under *Brown II,* however, "the apparently successful plaintiff in the *Brown* case got no more than a promise that, some time in the indefinite future, [she as well as other blacks] would be given the rights which the Court said [she] had."[21]

Some of the Justices, including the Chief Justice, later indicated that it had been a mistake to qualify desegregation enforcement by the phrase "all deliberate speed." Justice Black's son quotes him as saying, "It tells the enemies of the decision that for the present the status quo will do and gives them time to contrive devices to stall off segregation."[22] At the time of *Brown,* however, all the Justices thought that the Court should not require the South to swallow the desegregation pill all at once.[23] The Chief Justice himself had told a law clerk "that reasonable attempts to start the integration process is all the Court can expect in view of the scope of the problem, and that an order to immediately admit all negroes in white schools would be an absurdity because impossible to obey in many areas."[24] Even before the first *Brown* decision, Justice Black had warned the conference that enforcement would give rise to a "storm over this Court."[25]

The way to avoid such a storm, the Justice urged, was to allow time for enforcement: "Let it take time. It can't take too long."[26] *Brown II* was intended to secure the needed time by decentralizing enforcement.[27] The responsibility for implementing the *Brown I* principle was placed on the district courts, and the broad guides given them by *Brown II* left them with the widest discretion in the matter.

This approach did provide time for enforcement, but it also countenanced delay in vindicating constitutional rights. If, as Justice Black predicted at the *Brown II* conference, there was to be only "glacial movement" toward desegregation in the South,[28] that was true in large part because of the *Brown II* decision itself. "All deliberate speed" may never have been intended to mean indefinite delay, yet that is just what it did mean in much of the South—at least until the Supreme Court itself felt compelled to correct the situation, more than a decade after the *Brown* decisions.

Until the Supreme Court acted to clarify the meaning of *Brown,* the prevailing theme in desegregation enforcement was set in the lower federal courts. Most influential in this respect were opinions of Circuit Judge John J. Parker. Though his nomination to the Supreme Court in 1930 had been turned down by the Senate because he was supposedly "unfriendly" to labor and because of a statement against black "participation . . . in politics" made in the heat of a gubernatorial race,[29] by the time of *Brown* Parker had become one of the most respected federal judges in the country, whose decisions often pointed the way for other judges.

On remand of one of the cases decided in *Brown*, Judge Parker, writing for a three-judge district court, declared that *Brown*

> has not decided that the states must mix persons of different races in the schools or must require them to attend schools or must deprive them of the right of choosing the schools they attend. What it has decided, and all that it has decided, is that a state may not deny to any person on account of race the right to attend any school that it maintains. . . . The Constitution, in other words, does not require integration. It merely forbids discrimination.[30]

The distinction between "Thou Shall Not Segregate" and "Thou Shall Integrate" is of crucial importance.[31] Under the Parker interpretation, all that was required for compliance with *Brown* was to cease *governmental* action legally excluding black children from "white" public schools. But there was no duty to end existing dual school systems, provided only that they were no longer compelled by state law.

Brown itself was met in the South by various measures ranging from pressure and intimidation to school closings and the doctrine of "interposition," a doctrine that went back to pre–Civil War southern statesmen such as John C. Calhoun.[32] Less extreme were the pupil-placement statutes enacted in ten southern states.[33] They provided for assignment of children to schools on an individual basis, determined by consideration of the nonracial factors mentioned in the laws. Among the factors in the different state laws were: available room and teaching capacity in various schools; suitability of curricula for particular pupils; adequacy of a pupil's academic preparation for a particular school and curriculum; scholastic aptitude and intelligence, and psychological qualifications of the pupil; the pupil's home environment; morals, conduct, health, and personal standards of the pupil; as well as a catchall for "all relevant matters."[34] The statutes created a detailed administrative procedure that must be followed when a pupil was dissatisfied with a particular assignment.

The key factor in the pupil-placement laws was the need for individ-

ualized consideration of requests for transfers, as well as the detailed
provision of administrative procedures for those contesting assignments.

> In short, the statutes, functioning as intended, make mass integration al-
> most impossible, place the burden of altering the status quo upon indi-
> vidual Negro pupils and their parents, establish a procedure that is diffi-
> cult and time-consuming to complete, and prescribe standards so varied
> and vague that it is extremely difficult to establish that any individual
> denial is attributable to racial considerations.[35]

Despite this, the judicial tendency during the early post-*Brown* years
was to refuse to go beyond the nonracial face of the pupil-placement
laws. An early influential decision was that in *Carson* v. *Warlick*[36]—with
the opinion again written by Judge Parker. The Parker opinion pro-
nounced North Carolina's pupil-placement law constitutional

> upon its face. . . . Somebody must enroll the pupils in the schools.
> They cannot enroll themselves; and we can think of no one better quali-
> fied to undertake the task than the officials of the schools and the school
> boards having the schools in charge. It is to be presumed that these will
> obey the law, observe the standards prescribed by the legislature, and
> avoid the discrimination on account of race which the Constitution for-
> bids. Not until they have been applied to and have failed to give relief
> should the courts be asked to interfere in school administration.[37]

The Supreme Court declined to review this court of appeals decision.[38]

A year later, in *Shuttlesworth* v. *Birmingham Board of Education,*[39]
the highest Court went further and affirmed a decision refusing to strike
down the Alabama pupil-placement law. The lower court quoted the
Parker statement reproduced above in upholding "the constitutionality of
the law *upon its face.*" According to the court, "The School Placement
Law furnishes the legal machinery for an orderly administration of the
public schools in a constitutional manner by the admission of qualified
pupils upon a basis of individual merit without regard to their race or
color. We must presume that it will be so administered."[40]

The Supreme Court affirmed this decision, upholding the Alabama
law. It did so in a one-sentence per curiam, which laconically stated that
the Court was affirming "upon the limited ground upon which the District Court rested its decision."[41]

It soon became obvious that the pupil-placement laws were being used
to frustrate compliance with the *Brown I* principle. Thus, in Charlotte,
North Carolina—under the statute that had been given the Parker imprimatur—there were but three blacks in white schools in 1957, four in
1958, and only one in 1959.[42] In Alabama the situation was even worse.
During the lower-court argument in *Shuttlesworth,* counsel for plaintiff

stated, "I might mention this offhand, one Negro child was placed in a school designated as a White school in Alabama but he was removed three weeks hence because of community pressure and ill will generated by the pressure."[43] So far as is known, no other black had been assigned to attend the same school as a white under the Alabama law.[44]

By the 1960s, the federal courts were ready to recognize the pupil-placement laws for what they were, devices designed to maintain hard-core segregation, softened at best by a few token blacks in white schools.[45] A 1962 fourth circuit case involved a challenge to a pupil-placement law under which only nine black children were admitted to white schools in Roanoke, Virginia. Blacks denied admission sued to enjoin the school board from continuing their essentially segregated system. They sued even though they had not bothered to exhaust the administrative remedies under the placement law. The court did not require them to do so before granting relief.[46] As summarized by Alexander Bickel, "The pupil assignment practices of Roanoke were 'infected,' the court said, with racial discrimination. Such practices were tolerated in earlier cases 'as interim measures only,' but their day was now done."[47]

McNeese and Goss Cases

By this time the Supreme Court was also ready to intervene more actively in the desegregation process, for it had become apparent by then that the *Brown II* formula in practice meant indefinite delay in the implementation of the *Brown I* principle. Justice Black's prediction at the *Brown II* conference that there would be only "glacial movement" toward desegregation[48] had proved all too accurate. "Deliberate speed, as that term was used in *Brown II*, was never more than foot shuffling in the South. For most school districts it was an excuse to experiment a little with freedom of choice, to talk about building some new schools, to move a few bodies, to wonder about redrawing a zone line—to fiddle around and not do very much of anything."[49]

If the pace of *Brown* enforcement was to be speeded up, the Supreme Court itself would have to indicate that the time for delay was running out. The Court first did so in two cases decided on June 3, 1963. The first was *McNeese v. Board of Education*.[50] Unlike the other school cases decided by the Warren Court, *McNeese* did not arise in the South, but in Illinois. The claim was of segregation within a single school, with black students attending classes in one part of the school, separate and apart from whites, and compelled to use entrances and exits separate from the whites. The school board moved to dismiss on the ground that plaintiffs had not exhausted the administrative remedies provided by state law. The lower courts granted the motion, but the Supreme Court reversed, ruling that in such a case, where violation is alleged of the right vindicated in *Brown I*, plaintiffs need not exhaust unpromising state administrative remedies before bringing their action in a federal court.

The *McNeese* opinion does not refer at all to the pupil-placement laws. Indeed, only in the dissenting opinion of Justice Harlan are the cases under those laws (particularly *Carson* v. *Warlick*[51] and the Supreme Court's refusal to interfere with Judge Parker's ruling there) even cited. In these cases, according to Harlan, the federal courts followed the "wise approach" of requiring exhaustion.[52] By refusing to follow that approach in *McNeese*, the Court, Harlan argued, effectively repudiated the Parker ruling and delivered a fatal blow to the pupil-placement laws themselves.

In the second June 3, 1963, case, *Goss* v. *Board of Education*,[53] decided the same day as *McNeese*, the Court went out of its way to express dissatisfaction with the torpid pace of desegregation enforcement. The case arose out of transfer provisions in the desegregation plan adopted by the Knoxville, Tennessee, school board. The plan provided for school assignment on the basis of nonracial residential attendance zones, but permitted a pupil to transfer from a school where his or her race was in the minority or from a school previously serving the other race only. A unanimous Court struck down the transfer provisions, because race was the "absolute criterion for granting transfers which operate only in the direction of schools in which the transferee's race is in the majority." Race as "the factor upon which the transfer plans operate . . . is no less unconstitutional than its use for original admission or subsequent assignment to public schools." Equal protection is violated, since "racial segregation is the inevitable consequence" of the "obvious one-way operation" of the transfer provisions.[54]

The opinion of Justice Clark was careful to point out that transfer provisions not based upon race would present a different case: "This is not to say that appropriate transfer provisions, upon the parents' request, consistent with sound school administration and not based upon any state-imposed racial conditions, would fall."[55]

Justice Douglas had objected to this qualification in the *Goss* draft opinion. He circulated a draft concurrence noting that he did not join that part of the opinion containing this statement. The problem for him, wrote Douglas, was illustrated by the following hypothetical case: "Suppose an antisemitic parent files an application for transfer of his child from a school that is predominantly Jewish. The reason given is the child's adverse reaction to the school. The real reason is antisemitism. If the State bases its transfer system on those grounds, it acknowledges antisemitism as a legitimate ground for state action." The same would be true in comparable cases, for example, where "a Jewish parent could invoke state action to remove his child from a German school, a white from a Negro school, and a Negro from a white school, though racial segregation were not 'by its terms' the basis of the transfer plan."

According to the Douglas draft, "If a parent can get the State to transfer his child from one public school to another because of his or the child's racial prejudices, then the State is building its public school system on lines outlawed by *Brown*." In fact, the Justice asserted, "such a state transfer system indeed would undo what I thought we did in the *Brown* decision. If it were consistently applied we would shortly be back at the starting point of *Plessy v. Ferguson*."[56]

Despite the Douglas strictures, Justice Clark refused to remove the offending passage from the *Goss* opinion. The Chief Justice and the others were able to persuade Douglas that a unanimous opinion in a desegregation case was more important than his objection and the Justice withdrew his separate opinion.

In his draft *Goss* opinion, Justice Clark had stated that transfers might be permissible where the pupils' "emotional status" made them appropriate. This led Justice Brennan to write to Clark:

I was reading the other day of Judge Scarlett's decision in Georgia refusing to order the admission of Negro children to Georgia schools on the ground that the evidence adduced before him established that it was to the best interest of both races to keep colored and white children segregated.[57] I understand that this evidence by some psychologists from New York University was offered to prove that Gunnar Myrdal was wrong. Do you think there is any possibility that your first full paragraph on page 6, particularly the suggestion that transfers are appropriate where "emotional status" is involved, could possibly be interpreted as lending support to Judge Scarlett's view?[58]

The reference to "emotional status" was deleted from the final *Goss* opinion.

The *Goss* opinion referred to the difficulties that had led the Court to frame its *Brown II* mandate in terms of "all deliberate speed." But, it went on, "Now, however, eight years after this decree was rendered and over nine years after the first *Brown* decision, the context in which we must interpret and apply this language to plans for desegregation has been significantly altered."[59]

For the first time since *Brown* had begun to be implemented, the Court indicated that its patience was running out. The *Brown II* invitation to delay had given way to a need for more effective enforcement. As the Court put it in a case involving park segregation decided a week before *Goss:*

> Given the extended time which has elapsed, it is far from clear that the mandate of the second *Brown* decision requiring that segregation proceed with "all deliberate speed" would today be fully satisfied by types of plans or programs for desegregation of public educational facilities which eight years ago might have been deemed sufficient.[60]

Too Much Deliberation, Not Enough Speed

The next year the Supreme Court was confronted by the massive resistance of Prince Edward County, Virginia. The county had been the defendant in one of the original *Brown* desegregation cases. After the *Brown* decisions, the Supreme Court tells us, "Efforts to desegregate Prince Edward County's schools met with resistance."[61] In 1959, four years after *Brown II,* the fourth circuit court of appeals ordered the county to take immediate steps to desegregate its schools.[62] But the whites of Prince Edward, the site of the South's last strongholds at Petersburg and Appomattox, still rallied to the ghost of a brutal civil war, which with blurred, myth-befogged memory, they chose to recall as glorious. In their warped fervor, they saw themselves as the last stalwart hopes of a noble way of life that had, in fact, become a euphemism for shallow bigotry. Called by one commentator "America's most stubborn county,"[63] Prince Edward had resolved soon after *Brown* that it would never operate public schools "wherein white and colored children are taught together."[64]

After the court of appeals decree, the Board of Supervisors of Prince Edward County refused to levy any school taxes for the 1959–60 school year, explaining that they were "confronted with a court decree which requires the admission of white and colored children to all the schools of the county without regard to race or color." As a result, the county's public schools did not reopen in the fall of 1959 and remained closed until the Supreme Court decided the case five years later. State and county tuition grants and tax credits were provided for whites attending "private" schools, which excluded blacks.

The Justices were plainly affronted by the Prince Edward situation. A decade after *Brown,* here was one of the original *Brown* defendants, with its public schools shut down and only "private" white schools in operation. In Prince Edward at least, the black schoolchildren were not only not better off after *Brown,* but much worse; and the federal courts had been unable to help them during all that time.[65]

The Court's pique was evident throughout the opinion of Justice Black:

> The case has been delayed since 1951 by resistance at the state and county level, by legislation, and by lawsuits. The original plaintiffs have doubtless all passed high school age. There has been entirely too much deliberation and not enough speed in enforcing the constitutional rights which we held in *Brown v. Board of Education, supra,* had been denied Prince Edward County Negro children.[66]

In their decision, the Justices moved decisively to end Prince Edward's obstruction. The Court found flatly "that closing the Prince Edward schools and meanwhile contributing to the support of the private segregated white schools that took their place denied petitioners the equal protection of the laws."[67] In such a situation, the remedy required was far reaching. The district court was empowered not only to enjoin tuition grants and tax credits for the private schools, but also to "require the Supervisors to exercise the power that is theirs to levy taxes to raise funds adequate to reopen, operate, and maintain without racial discrimination a public school system in Prince Edward County like that operated in other counties in Virginia."[68]

The assertion of judicial authority to order the reopening and funding of the county's schools proved too much for Justices Clark and Harlan. For them, such an order went beyond the proper bounds of judicial power and they noted disagreement, without saying more, with the holding to that effect.[69] To the majority, however, the overriding consideration was "that relief needs to be quick and effective."[70]

The Black opinion concluded with tart language rejecting the *Brown II* formula: "The time for more 'deliberate speed' has run out, and that phrase can no longer justify denying these Prince Edward County school children their constitutional rights to an education equal to that afforded by the public schools in the other parts of Virginia."[71]

The same theme was repeated a year later in the 1965 decision in *Bradley v. Richmond School Board.*[72] It dealt with the issue of school faculty desegregation, an issue at least as touchy as that of desegregation of the pupils themselves. To many in the South, it was unthinkable that blacks would be permitted to discipline their offspring and teach them "nigra" talk.[73] By the time of the *Bradley* case, in fact, not one of the 36,500 black teachers in Alabama, Louisiana, or Mississippi taught with any of the 65,400 whites.[74]

In *Bradley,* the Court forbade further delay in ending teacher segregation. The lower court had approved two school desegregation plans without any hearing on the claim that faculty segregation made the plans inadequate. The Supreme Court granted certiorari and reversed in a short per curiam opinion, holding that petitioners were entitled to full hearings upon their contention. Justice Brennan, who wrote the per curiam,[75] stated the principle that school desegregation plans might not be approved "without considering, at a full evidentiary hearing, the impact on those plans of faculty allocation on an alleged racial basis."[76]

Justice Black had urged that the opinion stress the point he had made in the *Prince Edward* case, that desegregation must now occur immediately, without further delay.[77] The per curiam did so in language comparable to that used by Black in *Prince Edward*:[78] "more than a decade has passed since we directed desegregation of public school facilities 'with all deliberate speed.' . . . Delays in desegregating school systems are no longer tolerable."[79]

The Changing Traffic Light: From *Brown* to *Green*

After *Brown* itself, the key Warren Court school desegregation case was *Green v. County School Board*.[80] At issue in it was a so-called freedom of choice plan, allowing a pupil to choose his or her own public school, adopted by a Virginia school board. Freedom of choice plans, which were the converse of the pupil-placement laws under which assignment of pupils was made by the school board, had become widespread.[81] The vast majority of southern school districts that submitted desegregation plans under the guidelines issued by the Department of Health, Education and Welfare chose freedom of choice plans.[82] When *Green* was decided, most southern school districts operated under such plans."[83]

In his opinion, already referred to, on remand of one of the original cases joined in *Brown,* Circuit Judge Parker asserted, "Nothing in the Constitution or in the decision of the Supreme Court takes away from the people freedom to choose the schools they attend."[84] The difficulty was, however, that "freedom of choice" was, in practice, a tool for continuation of the southern status quo. As Judge McMillan put it, " 'Freedom of choice' was a device to *prserve* [sic] segregation. It did not aid in *eliminating* segregation."[85] Freedom of choice at best produced only token integration; as such, it only scratched the surface of the existing school system. "In most instances," writes one observer, "the term 'freedom of choice' is an unfortunate misnomer since forces acting on the Negro parents and their children, both subtle and overt, prevent the true exercise of a 'free choice.' "[86]

The reality under the freedom of choice plans is shown by the fact pattern in the *Green* case itself. It arose in a rural Virginia county in which blacks comprised some 57 percent of the school population. The county had two schools; one had been a white school, the other had been the school for blacks. For a decade after *Brown*, the situation remained unchanged. In 1965, in order to remain eligible for federal funds, the county adopted a freedom of choice plan. Under it, each pupil might choose each year between the two schools. If no choice was made, they were assigned to the school previously attended. In the three years in which the plan was in operation, not a single white child chose to attend the black school, and 85 percent of the black children in the county still attended the all-black school.

According to its counsel, however, the county had done all that could be expected under the circumstances. The attorney stressed that *Brown* had never required compulsory integration; it only required the states to "take down the fence" keeping pupils apart. To this, Chief Justice Warren interposed, "Isn't the net result that while they took down the fence, they put booby traps in the place of it, so there won't be any white children going to a Negro school? . . . Isn't the experience of three years . . . some indication that it was designed for the purpose of having a booby trap there for them, that they didn't dare to go over?"[87]

With Justice Marshall not voting, all the other Justices (except Harlan) voted to grant certiorari in *Green*. At the conference on the case, the Chief Justice took the same approach as he had at oral argument. He again stated the position that "freedom of choice is violative of constitutional rights." As he saw it, "the purpose of *Brown* was to break down segregation" and the county's plan failed to do that.[88] Except for Justice Black, who voted to affirm the decision upholding freedom of choice, the remaining Justices agreed with Warren that the plan, as it operated, invalidly maintained a segregated school system. Ultimately, Justice Black agreed to go along with the majority, enabling the Warren Court to maintain its record of unanimity on the merits in school desegregation cases.

The *Green* opinion was written by Justice Brennan, who had expressed himself at the conference as strongly as had the Chief Justice against the freedom of choice plan. This was the first time that Justice Brennan was given the opportunity to write an opinion in his own name in a school case (though he had written the opinion in *Cooper* v. *Aaron*[89]—the Little Rock school case—and the *Bradley* teacher desegregation case,[90] the opinions had not come down as Brennan opinions). The Justice decided to write a *Green* opinion that would deal with the problem as forcefully as possible and, at the same time, be modeled on the *Brown I* opinion itself, "short, pungent, and to the point."[91] As told by his clerks, Brennan wanted "to sweep away much of the dogma that has grown up in *de jure* litigation to hinder desegregation efforts: the

Briggs dictum [of Judge Parker], the semantic distinction between 'desegregation' and 'integration' [to stress] the duty of school boards to maximize integration where feasible." In sum, the clerks say, "what the Justice wanted to do was to arm the Civil Rights Division and H.E.W., plus the 'right' courts of appeals and judges, with the ammunition to go after the recalcitrants."

Justice Brennan circulated a strong draft *Green* opinion. As the Chief Justice had done in *Brown*, the Justice emphasized the "inferiority" of education in segregated schools:

> The stigma of inferiority which attaches to Negro children in a dual system originates with the State that creates and maintains such a system. So long as the racial identity ingrained, as in New Kent County, by years of discrimination remains in the system, the stigma of inferiority likewise remains. Only by reorganizing the system—extending to pupils, teachers, staff, facilities, school transportation systems, and other school-related activities—can the State redress the wrong of depriving Negro school children "of some of the benefits they would receive in a racial[ly] integrated school system." It was to that end that *Brown II* commanded school boards to bend their efforts.[92]

However, Justice Brennan recognized the need to continue the tradition of unanimity in school segregation cases. The *Green* draft was circulated on May 16, 1968—the eve of the fourteenth anniversary of the first *Brown* decision. As the conference the next day, Justices White and Harlan advised Justice Brennan that they could not join the opinion so long as it contained the references to "inferiority." They said that the references risked angering desegregation opponents even further and rendering compliance even harder to obtain. Justice White also asserted that in his view modern sociological and psychological data did not support the notion of "stigma" relied on by the *Brown I* opinion. Had he been on the Court, White said, he would not have agreed with *Brown I*'s famous footnote 11.[93] Both Justices White and Harlan urged Justice Brennan to avoid raising the footnote 11 controversy again and to leave the notion of stigma "implicit" in the opinion.[94]

After consulting the other members of the Court (particularly the Chief Justice, who had been reluctant to have anyone "fooling around" with the *Brown* decisions[95]), Justice Brennan agreed to delete the "inferiority" language and the passage quoted above does not appear in the final *Green* opinion. Justice Brennan also made some changes in the draft which were suggested by Justice Black and enabled the latter to join the opinion despite his original dissenting vote.[96] The result was a unanimous *Green* opinion striking down freedom of choice plans.

Though, as just seen, Justice Brennan had to tone down his *Green* opinion, the opinion as issued was still unusually forceful—indeed, the strongest Supreme Court opinion on the subject since *Brown*. The key factor requiring the freedom of choice plan to be invalidated was its failure to bring about elimination of the "dual [school] system, part 'white' and part 'Negro.' It was such dual systems that 14 years ago *Brown I* held unconstitutional and a year later *Brown II* held must be abolished."[97]

The *Green* opinion declares that the question for the court in a school segregation case is whether the school authorities have "achieved the 'racially nondiscriminatory school system' *Brown II* held must be effectuated." It was not enough for school boards, such as that in the Virginia county before the Court, merely to remove the legal prohibitions against black attendance in white schools. Instead, "School boards such as the respondent then operating state-compelled dual systems were . . . clearly charged with the affirmative duty to take whatever steps might be necessary to convert to a unitary system in which racial discrimination would be eliminated root and branch."[98]

The Court again stressed that it was no longer willing to countenance the South's "deliberate perpetuation of the unconstitutional dual system." Delays "are no longer tolerable" and the *Prince Edward* statement, "The time for mere 'deliberate speed' has run out," was repeated. The courts were no longer to tolerate school plans that might bring about desegregation at some future time. On the contrary, "The burden on a school board today is to come forward with a plan that promises realistically to work, and promises realistically to work *now*."[99]

Nor was it enough for the lower courts to find that a given school board had met this burden. The judicial responsibility did not end with approval of a plan that discharged the school authority's affirmative duty to convert to a unitary system. The plan must also "prove itself in operation." Thus, "whatever plan is adopted will require evaluation in practice, and the court should retain jurisdiction until it is clear that state-imposed segregation has been completely removed."[100] Under *Green*, the district courts were now expressly vested with the affirmative duty to supervise the operation of desegregation plans. The clear implication was that they should do whatever they deemed necessary to ensure that those plans proved effective in practice. Discharge of the judicial duty here might well involve the courts in the intimate details of school administration. Perhaps, under the *Brown II* approach, the Supreme Court itself avoided the danger of acting as what Justice Frankfurter had termed "a super-school board."[101] But the same would not necessarily be true of the district courts now charged with the affirmative mandate imposed by the *Green* opinion.

After the decision striking down freedom of choice plans was announced, Governor Lester Maddox of Georgia acknowledged it by ordering all state flags flown at half-mast. The Justices also understood the significance of their decision. When Chief Justice Warren joined the *Green* opinion, he wrote Justice Brennan, "When this opinion is handed down, the traffic light will have changed from *Brown* to *Green*. Amen!"[102]

Montgomery County and Racial Balance

The last Warren Court school case, *United States* v. *Montgomery County Board of Education*,[103] was decided toward the end of the Chief Justice's final term. In it, District Judge Frank Johnson had issued an extensive desegregation order for the schools in Montgomery County, Alabama. Nearly all aspects of the order were accepted by the school board. The board did, however, challenge the part of the order that dealt with faculty desegregation. Judge Johnson had found "that the teachers are assigned according to race; Negro teachers are assigned only to schools attended by Negro students and white teachers are assigned only to schools attended by white students." The judge's order provided that the school board must move toward a goal under which "in each school the ratio of white to Negro faculty members is substantially the same as it is throughout the system."[104] In addition, the order set forth a specific schedule for attainment of a desegregated faculty. As a first step, each school was to have at least two teachers of each race, and schools with twelve or more teachers were to have at least one black for every five whites or vice versa.

The court of appeals reversed Judge Johnson's faculty desegregation order. Viewing the order as requiring "fixed mathematical" ratios, it held that "compliance should not be decided solely by whether [the school board] has achieved the requisite numerical ratios." The part of the order setting the goal for the next year should be modified to require only *"substantially* or *approximately"* Judge Johnson's five-to-one ratio.[105]

All the Justices voted in favor of granting certiorari and all voted to reverse at the May 2, 1969, conference following the argument of the case. The opinion was assigned to Justice Black, who soon prepared a draft opinion, which he circulated on May 20. The draft was quickly approved by the others, though Justice Harlan did suggest minor changes. The most significant of these was the omission of the sentence "It is possible that we may be in error in believing it best to leave Judge Johnson's order as written rather than as modified."[106] A Harlan letter to Justice Black stated, "While I think I understand its purpose, particularly in the context of what next follows, the sentence might carry to others a quite different flavor." Harlan also suggested removing the phrase that the

Court was "accepting the choice of the judge on the spot." That phrase, Harlan wrote, "might prove embarrassing to the Court of Appeals in controlling District Judges who are not Frank Johnsons, and this certainly would be unfortunate."[107]

Aside from the minor changes suggested by Justice Harlan, the final *Montgomery County* opinion was virtually the same as the Black first draft. In some ways it went even further than *Green.* The Black opinion expressly affirmed the shift from the *Brown* prohibition against state-compelled segregation to the affirmative duty "of insuring the achievement of complete integration at the earliest practicable date."[108]

Green had used the statistics on racial composition of schools to invalidate the school board's freedom of choice plan. Now Judge Johnson, having found that previous plans had not produced integrated faculties, had imposed his own quantitative standards that the board was to meet. And the Supreme Court expressly affirmed his requirement: "Judge Johnson's order now before us was adopted in the spirit of this Court's opinion in *Green* v. *County School Board* . . . in that his plan 'promises realistically to work, and promises realistically to work *now.*' "[109] For the first time, the Supreme Court sustained the inclusion of affirmative numerical goals (the court of appeals had treated them as racial quotas) in a school desegregation decree.[110]

Did the Court's approval of racial ratios in faculty assignments in *Montgomery County* imply approval of a similar requirement for students as well?[111] As we shall see, Chief Justice Burger in his first draft *Swann* opinions answered this question in the negative and limited *Montgomery County* to teachers, saying it did not apply to student assignments. The other Justices, however, disagreed and prevailed upon the Chief Justice to broaden his restricted interpretation. In the final *Swann* opinion, the Court treats the Johnson order as a precedent for Judge McMillan's use of racial ratios, affirming expressly: "The principles of *Montgomery* have been properly followed by the District Court and the Court of Appeals in this case."[112] *Montgomery County* was thus, like *Green,* a direct precedent for the *Swann* decision.

School Desegregation and the Warren Court

The Warren Court decisions on school segregation involve a progression from *Brown* to *Green* and *Montgomery County,* in which the Supreme Court itself plays an increasingly active role. The foundation of the law in this area was, of course, laid down in *Brown* v. *Board of Education.* In its *Brown I* decision, the Court ruled that segregation in public schools violated the Constitution. *Brown II* provided for implementation of the

Brown I principle by the district courts, whose enforcement discretion was limited only by the broad guidelines contained in the *Brown II* opinion.

During the nearly ten years that followed, the Supreme Court left the enforcement problem to the lower courts and all but kept its hands off school cases. "For a decade after *Brown,* the most remarkable thing about the Supreme Court's action with respect to school segregation is that it acted so little. Having declared the war, it largely withdrew from the battle."[113]

By 1963, however, it had become apparent that the war against school segregation was not being won, in large part because of the Supreme Court's post-*Brown* failure to assert an effective leadership role. In *Montgomery County,* the Court noted that *Brown II* had been unwilling to leave the "responsibility for abolishing the system of segregated schools in the unsupervised hands of local school authorities, trained as most would be under the old laws and practices, with loyalties to the system of separate white and Negro schools." Instead, "The problem of delays by local school authorities during the transition period was therefore to be the responsibility of courts, local courts so far as practicable, those courts to be guided by traditional equitable flexibility to shape remedies in order to adjust and reconcile public and private needs."[114]

The difficulty was that many of the lower federal courts had not moved aggressively enough to eliminate dual school systems. The result was that, in all too many areas, the schools still "operated, so far as actual racial integration was concerned, as though our *Brown* cases had never been decided."[115] If this situation was to be corrected, the "coercive assistance" of the Supreme Court itself "was imperatively called for."[116]

That the Court began to take an increasingly active role in the school cases, starting with the *McNeese* and *Goss* cases almost a decade after *Brown I,* may be explained by the Justices' increasing exasperation at southern refusals to implement *Brown.* As their irritation grew during the next five years, so did their intervention. In an interview not long before his death, Justice Clark (who had participated in all the Warren Court cases before *Green*) noted that judicial power in the school cases "grew in small individual steps but 'like Topsy' with no grand design." The Court, Clark said, had tried "to give localities a chance to make a change in their own ways." However, when confronted in case after case with "wholesale obstruction," it had no choice but to broaden federal judicial supervision—and, finally, in the *Swann* case, after Chief Justice Warren had retired, to order busing.[117]

The Warren Court cases support the Clark summary. From *Brown II* to *Goss,* the Court followed its hands-off policy, trusting to the school authorities and district courts to ensure elimination of school segregation. During the next five years, the Justices moved from their immediate post-*Brown* inaction to an increasingly activist stance, culminating in *Green,* the key Warren Court case after *Brown* itself.

Yet *Green* was more than an assertion by the Court of an active role in the process of implementing *Brown.* "Although not generally realized at the time," wrote one commentator, "the Supreme Court's 1968 decision in *Green v. County School Board* . . . worked a revolution in the law of school segregation comparable to, indeed more drastic than, that effected in *Brown.*"[118] This statement surely goes too far. But *Green* did alter the focus of inquiry in the school cases. In the process the Court changed the constitutional rule from the *Brown* prohibition against compelled segregation to an affirmative duty immediately to dismantle all dual school systems—a duty to be assumed by the federal courts where local authorities did not "come forward with a plan that . . . promises realistically to work *now.*"[119]

Green fixed the pattern for judicial enforcement in school cases and, as such, set the stage for the later decisions on the subject, including that in the *Swann* case. In the first place, *Green* drastically lessened the burden on plaintiffs seeking judicial relief. All that they would have to show was the existence of dual schools in the districts concerned. This was done in *Green* by showing that there was still an all-black school in the county attended by 85 percent of the black children. Whatever plan the county may have had to end segregation, it plainly was not working when so large a proportion of blacks remained segregated. The statistical imbalance alone demonstrated the existence of an invalid dual system.[120]

The showing of the existence of dual schools was all that was required under *Green.* The focus of inquiry now shifted. As the Chief Justice expressed it at the *Green* conference, "this is a problem of remedy and nothing more."[121] Once the dual system was demonstrated, the only question for the Court was that of remedy—what measures to take that would eliminate the dual system. The remedial starting point was that any plan adopted was to "take whatever steps might be necessary to convert to a unitary system in which racial discrimination would be eliminated root and branch."[122] The school board concerned had the burden of presenting a plan to accomplish this. However, if the board did not come forward with a plan that "promises realistically to work *now,*"[123] the implied alternative was that the district court should do so. From the *Brown* invalidation of prohibitions against blacks attending white schools, the Warren Court had moved to the *Green* affirmative duty to provide a fully integrated school system, with the federal judges having the ultimate responsibility for ensuring that the conversion from a dual to a unitary system took place as soon as possible.

The clear implication was that the district courts should issue whatever orders were necessary to bring about an integrated system, even if that required attention to administrative details that were normally within the school board's province. The Warren Court did not specify the remedial measures that might be ordered, though it did offer two tentative suggestions in a *Green* footnote. The first was geographical zoning: this "could be readily achieved . . . simply by assigning students living in the eastern half of the county to the New Kent School and those living in the western half of the county to the Watkins School." The second was pairing of the two schools in the county: "the Board could consolidate the two schools, one site (e.g., Watkins) serving grades 1–7 and the other (e.g., New Kent) serving grades 8–12."[124]

These remedial suggestions were anything but drastic; the first merely provided for attendance under a modified "neighborhood school" concept. But that was only because the county involved was a rural one with little residential segregation. The same suggestion would prove inadequate in urban areas with entirely different residential patterns. In such areas, would neighborhood schools be sufficient to meet the constitutional standard or would more drastic remedies be appropriate? Could those remedies include busing between white and black areas if the district court determined that it was needed to ensure meaningful integration?[125]

The Warren Court did not have to deal with these questions. But the *Green* opinion surely contains implicit answers. *Green* imposes the duty to take whatever steps are necessary to convert to a unitary school system, including measures that promise realistically to work—and to work *now*. In the typical urban school district, pupil assignments only to neighborhood schools would scarcely dismantle dual systems. Instead, they would merely mirror existing residential patterns. If only transportation of children to schools outside their neighborhoods would hold any promise of realistically working to end dual systems, *Green* clearly implies that such transportation may be ordered. From this point of view Justice Brennan was correct when he wrote to Chief Justice Burger, in a memorandum on the *Swann* case, that "all that we are really required to do here is fill in the outline constructed by *Green*."[126]

Green is thus the connecting case between the *Brown* and *Swann* cases. The Warren Court did not answer the question of school busing raised in *Swann* because the question was never presented to it. In light of the strong language in the *Green* opinion, however, can there be any doubt about how that Court would have answered it?

4

Burger's First
Desegregation Case

Swann was not the Burger Court's first school desegregation case. Soon
after the new Chief Justice was sworn in in 1969, his Court was faced
with *Alexander v. Holmes County Board of Education*,[1] the most important case of the first Burger term.

Though fifteen years had passed since *Brown I*[2] had ruled school
segregation invalid, Mississippi still "had in effect what is called a dual
system of public schools, one system for white students only and one system for Negro students only."[3] On July 3, 1969, the U.S. Court of Appeals for the Fifth Circuit entered an order directing the Department of
Health, Education and Welfare to submit desegregation plans for 33
Mississippi school districts to be put into effect at the beginning of the
new school year in September.[4] In August, however, both HEW and the
Department of Justice filed a motion for a delay until December 1. It
was the first time the federal government had favored a delay in a desegregation case, but the government's posture reflected the changed desegregation policy of the new Nixon administration. On August 28 the
court of appeals granted the government's motion; it suspended its July
3 order and postponed the date for submission of the new desegregation
plans to December 1.[5]

The black plaintiffs in the case then asked Justice Black, as circuit justice with supervisory authority over the fifth circuit court of appeals, to
vacate that court's suspension of the July 3 order. The petition to prevent
delay of desegregation in the Mississippi school districts presented Justice
Black with a dilemma. On the one hand, the Justice had been dismayed at
the massive resistance in the South to the desegregation requirement. Being from the deep South, Black had realized more than any other member

of the Court that, as he had put it at the *Brown II* conference, "the South would never be a willing party to Negroes and whites going to school together." Instead, he had said then that there would at best be only "glacial movement" toward desegregation.[6] But even Black did not foresee that there would have been *no* movement at all in a state like Mississippi a decade and a half after *Brown*. As the years since *Brown* went on, the Justice was less and less willing to countenance delay. His opinion in the *Prince Edward*[7] case showed that his patience with "all deliberate speed" had run out. So far as his own views were concerned, Black would have unhesitatingly granted the petition to prevent the delay permitted by the court of appeals.

At the same time, Black said that "my views as stated above go beyond anything this Court has expressly held to date." Even *Green* "might be interpreted as approving a 'transition period' during which federal courts would continue to supervise the passage of the Southern schools from dual to unitary systems."[8] It would be safer to have the whole Court decide the matter than to have one Justice, acting on his authority alone, attempt finally to lay to rest the ghost of the "all deliberate speed" formula.

In an in-chambers decision on September 5, Black refused to overturn the delay given by the court of appeals, "deplorable as it is to me." The opinion which the Justice issued did, however, clearly indicate his own dismay at the delay. As far as he was concerned, Black wrote, he would make the desegregation orders "effective not only promptly but at once—*now*" and "do away with that phrase [i.e., 'all deliberate speed'] completely."[9]

In his opinion Black expressed the hope that his decision would not be the last word in the matter: "I hope these applicants will present the issue to the full Court at the earliest possible opportunity."[10] The plaintiffs accepted this invitation and petitioned the Court for certiorari, which was granted unanimously.[11] The case was argued on Thursday afternoon, October 23. All the Justices were present except Justice Brennan, who did not sit because his wife was undergoing surgery that day.

The ornate courtroom, with its marble walls and stately side columns, was packed for the argument. The case assumed added significance both because it was the first major case under the new Chief Justice and because of the changed position of the Nixon administration. For the first time the government was arguing before the Court in support of southern attempts to delay desegregation.

The mood of the Justices during the argument was caught in a slip of the tongue by Justice Black. While questioning an attorney about the possible form of an order saying "that the dual system is over," Black asked what its nature should be "when we issue an order." Then, realizing the compromising implications of his use of the word "when," the Justice added, "If we do."[12] There was a flurry of laughter throughout the courtroom.

Conference

The Justices met for their usual Friday conference the next day, October 24. It is the normal Supreme Court practice for the Chief Justice to start the conference discussion by summarizing the issues and giving his views on the case. The others follow in order of seniority, starting with the senior Associate Justice and ending with the junior—that is, the Justice most recently appointed. In *Alexander,* however, there was a departure from the usual practice. The Chief Justice did not open the conference by stating his views. Instead, he recognized Justice Black, who began the discussion of the case. Speaking at some length, the Justice reviewed the origin and history of the phrase "all deliberate speed." Black stated that he had opposed its use in the second *Brown* opinion[13] because he had thought it would breed resistance to desegregation; he had, however, acquiesced in the *Brown II* opinion in order to preserve a unanimous Court. Black was emphatic in his assertion that he would never again join an opinion using the phrase. The Black position was that desegregation must come immediately, without the slightest delay or any ifs, ands, or buts.

After Black had finished, Justice Douglas spoke briefly; he stated his general agreement with the Black presentation, though indicating his willingness to be somewhat more flexible. Justice Brennan took the same approach when his turn came in the conference discussion. He said his position had been spelled out in an opinion he had issued in August 1969, as a circuit justice.[14] In that opinion, Brennan had struck down an order by the tenth circuit court of appeals, which had stayed immediate implementation of desegregation plans in Denver.

Before Brennan spoke, a more cautious view had been expressed by Justice Harlan. He agreed that the court of appeals order should be overturned, but he said that he favored taking account of some of the practical difficulties involved in immediate desegregation. Harlan concluded that it was not enough for the Court to order desegregation "now." At least some time had to be given for the Court's order to be carried out. The other Justices who spoke—Stewart, White, and (somewhat surprisingly) Marshall—expressed agreement with Harlan's more restrained approach.

All the Justices who expressed a view at the October 24 conference agreed that the court of appeals should be reversed. But there were three positions expressed on the amount of time to be allowed to carry out the Court's order. At one extreme was Justice Black, who wanted immediate desegregation. He was unwilling to allow any language at all in the opinion that might give the Mississippi school boards any pretext either to delay outright or return to court to request more time. Justice Douglas and Brennan were less inflexible on this point; both were in favor of rapid desegregation, but were willing to grant a limited period in which it might be achieved. Thus, Justice Brennan spoke of giving the school boards eight days to implement whatever plans the court of appeals approved; the latter, in turn, was to have three days from the time the Supreme Court's judgment issued to devise the plans. The other Justices felt that the boards had to be given more time to act and that the opinion should so provide.

At the end of the conference discussion, the Chief Justice again declined to express his views, but he said that he would prepare a draft opinion for the Court. As was seen in Chapter 2, the Supreme Court practice is for the Chief Justice to assign opinions only when he is part of the conference majority. Burger alone had not stated his views on the case at the *Alexander* conference. Despite this, he had exercised the assigning power to take the opinion himself. He told the conference that he would circulate a draft to serve as a basis for discussion. The others made no objection despite the apparent violation of the practice on opinion assignments.

Burger Draft Order

The Chief Justice had a two-page printed "Proposed Order and Judgment" ready the next day, October 25. Because it was a Saturday, the draft was circulated to the Justices' homes, together with a *Memorandum to the Conference,* headed "Confidential."[15] The memo stated, "Justice Harlan, Justice White and I met today . . . and developed the enclosed order to be followed by an opinion." The draft itself, according to Burger, "reflects not necessarily our final view but a 'passable' solution of the problem."

On the question of time—the key issue—Burger wrote, "We have concluded, tentatively, to avoid fixing any 'outside' date. I am partly persuaded to do this because of the risk that it could have overtones which might seem to invite dilatory tactics."

The Burger draft began with an introductory paragraph that declared, "The question presented is one of paramount importance, involving as it does the continued denial of fundamental rights of some 137,000 school children—Negro and White—who are presently attending Mississippi schools under segregated conditions notwithstanding and in violation of numerous pronouncements of this Court from May 17, 1954, to date." It went on to say, "Because of the gravity of the issues and the exigency of prompt compliance with the Constitution, we deem it appropriate to enter the following order, based on our review of the submissions and consideration of oral argument, with opinion to follow this judgment."

The second paragraph led directly to the order itself:

> The petitioners having urged immediate termination of a dual school system based on race and the Attorney General having urged that result without awaiting the beginning of the 1970–1971 school year,
> *It is hereby adjudged, ordered and decreed.*

The order itself, on two printed pages, contained five numbered paragraphs. The first vacated the court of appeals delaying order and remanded the case to that court "for a determination forthwith" whether the plans submitted by HEW "are reasonable and adequate interim means to achieve immediate termination of any system of dual schools based on race or color."

The order's second paragraph read: "The Court of Appeals shall enter its order for interim relief on or before November 10 (?) 17 (?) 1969." The next paragraph provided that all further proceedings in the case "shall be conducted on an expedited schedule consistent with the urgency of these cases."

The fourth paragraph stated that the order of the court of appeals "for interim relief shall be made operative by said Court at the earliest possible time and date after the order of said Court." The order's last paragraph provided that the Supreme Court's mandate in the case "shall issue forthwith and the Court of Appeals is directed to lay aside all other business of the Court to carry out this mandate."

There is a version of the draft Burger order in Justice Harlan's papers, headed "Mr. Justice Harlan's Submission of October 25, 1969: Proposed Order and Judgment," which indicates that the first draft of the order circulated by the Chief Justice was prepared by Harlan. Despite this, it may be doubted that Harlan wholly approved of the draft as it was circulated on October 25. This was shown by the Justice's later strong objection to the assertion in the second paragraph of the Burger draft order's preamble that the attorney general had urged that no desegregation delay should be permitted.

In a letter to the Chief Justice on October 28, Harlan referred to this paragraph as "indicating in effect that the Government is in accord with the accelerated desegregation program which our order envisages." Harlan wrote:

I think such an intimation is quite unpersuasive because, although the Government did envisage the accomplishment of steps toward desegregation prior to the commencement of the school year 1970–71, it has continued to maintain the proposition that the HEW should be given until December 1 to file its plans. Frankly, I think it undesirable to blink the fact that the Government stands in opposition to the central and only issue in the case before us.

Brennan Redraft

When he read the Burger draft order, Justice Brennan concluded that it was unsatisfactory. The next morning, Sunday, October 26, he telephoned Justice Black's home, but learned that he had gone to his chambers to prepare a dissent. Brennan then drafted a revised version of the Burger order. The Brennan draft, written down by the Justice himself, bypassed the Burger preamble and went directly to the order itself, beginning with, "It is hereby adjudged, ordered and decreed."

The Brennan version, like the Burger draft, contained five numbered paragraphs, but the tone and substance were entirely different. Brennan's first paragraph sounded the final death knell of the "all deliberate speed" formula: "Desegregation of segregated school systems according to the standard of 'all deliberate speed' is no longer constitutionally permissible. The obligation of the federal courts is to achieve desegregation of such systems now." This Brennan language was an improved version of the Justice's opinion in *Green v. County School Board*;[16] it was to form the basis of the key passage in the final *Alexander* opinion.

In Justice Brennan's second paragraph, the court of appeals August 28 order was vacated and the court was directed "to enter an order to effect immediate termination of the unconstitutional dual school systems." The court of appeals might order the HEW recommendations into effect, with necessary modifications, "provided the Court determine forthwith (that is, not later than October 30 next) that such recommendations as so amended or modified suffice to achieve immediate termination of any system of dual schools based on race or color." The October 30 date was based on the fact that Brennan expected the Supreme Court to issue its order on the following day, October 27.

The third Brennan paragraph stated that the court of appeals might make its determination and order "without further arguments or submissions." Paragraph four provided that, while the court of appeals order was being carried out, the district court might consider objections and amendments, but "the court of appeals order shall not in any manner or term be suspended pending decision on such exceptions or amendments."

The last paragraph that Brennan wrote read: "5. As is"—meaning that no changes were made in paragraph five of the Burger draft order.

The draft order that Justice Brennan jotted down after he had read the Chief Justice's proposed order was similar in most respects to the final *Alexander* Court order. Two crucial changes had been made by the Justice. In the first place, the Brennan draft expressly asserted the demise of the "all deliberate speed" formula. Though this passage was taken out of the final *Alexander* order, the substance of Brennan's language on the point was to be included in the preamble of the per curiam opinion that the Court delivered in the case. As already stated, this was to be the key passage in the *Alexander* opinion. In addition, in the Brennan draft, all the references in the Burger order to "interim" relief and relief "at the earliest possible time" were replaced by references to "immediate" relief or relief "forthwith" (though the court of appeals was given up to three days to consider and modify the HEW recommendations).

Black Dissent

At about noon on Sunday, October 26, after he had written out his draft order, Justice Brennan telephoned the Black chambers. He was told that Black had already prepared a dissent. Brennan then read the text of his proposed order to Justice Black's secretary. As it turned out, though his call came too late to head off the Black draft, Brennan's order did substantially influence the dissent that Black circulated.

The Black draft dissent was circulated to the homes of the Justices during the afternoon of October 26, together with a covering one-page typed *Memorandum to the Members of the Conference*. The Black memo informed the Justices that if the Burger draft obtained a majority and the Court wanted to issue the Chief Justice's order the next day, "I would not want to delay such action, but will dissent as I have in the opinion circulated herewith."

The Black memo went on to state, "While a dissent at this time may seem premature, this procedure has been followed only to avoid further delay." This was, of course, not the real reason for Black's circulation of his draft. The Justice acted as he did because he knew that the threat of his dissent in a school segregation case would induce the others to agree to the decision that he desired, one that would order immediate desegregation with no ifs, ands, or buts.

Black's covering memo concluded: "One more thought . . . about the Court's suggestion that a Court opinion will later follow this order. I am opposed to that. There has already been too much writing and not enough action in this field. Writing breeds more writing, and more disagreements, all of which inevitably delay action." This was followed by a typical Black peroration: "The duty of this Court and of the others is too simple to require perpetual litigation and deliberation. That duty is to extirpate all racial discrimination from our system of public schools NOW."

The draft dissent which was circulated with this Black memo consisted of seven typed legal-size pages. Headed "MR. JUSTICE BLACK, dissenting," it began with a statement of the *Brown* decisions and a strong attack on "all deliberate speed." According to the draft, "This phrase 'all deliberate speed,' apparently was casually picked up from an opinion of Mr. Justice Holmes in *West Virginia* v. *Virginia* [*sic*], 222 U.S. 17, a case which did not involve federal constitutional rights. As Mr. Justice Holmes recognized, the phrase connotes delay, not speed, and its use in *Brown II* with reference to delay in enforcement of cherished constitutional rights has proven an unfortunate one."

But even the *Brown II* use of the phrase, Black went on, was plainly never intended to permit the fifteen-year delay in desegregation that had occurred: "It is almost beyond belief that the factors mentioned by this Court in *Brown II* to permit some slight delay in 1954 are precisely the same considerations relied upon in this case to justify yet another delay in 1969."

"My dissent to the Court's order," the Black draft declared, was based on the fact that

> that order revitalizes the doctrine of "all deliberate speed" under the repeated euphemisms of "interim order" and "interim relief" to be made "operative" at "the earliest possible time," in spite of the fact that we have already emphatically repudiated the "deliberate speed" delay formula at least twice.[17] Any talk of "interim" orders necessarily implies that complete, total and immediate abolition of the dual school system need not come about, and the phrase "the earliest possible time" is ominously reminiscent of the phrase "as soon as practicable" used in *Brown II*.

The Black draft referred to the lower court's finding that disastrous educational consequences might follow from immediate desegregation. "In my opinion," the draft asserted, "there can be no more disastrous educational consequence than the continuance for one more day of an unconstitutional dual school system such as those in this case."

The strong Black conclusion was:

> The time has passed for "plans" and promises to desegregate. The Court's order here, however, seems to be written on the premise that schools can dally along with still more and more plans. The time for such delay, I repeat, we have already declared to be gone. On the basis of prior hold-

ings the States and the Nation have nothing now to deliberate about further.

On the contrary, it must be recognized "from this moment on that a separate school system for whites and blacks cannot longer exist in this country, and to make this recognition a reality, I would have the Court issue the following order."

The draft dissent then ended by quoting almost verbatim the draft order which Justice Brennan had read to Black's secretary a few hours earlier. The only changes of importance were in the second paragraph. Black had omitted the reference to October 30 as the latest date for court of appeals action. The Black paragraph contained no reference to any date. Instead, Black required the court of appeals

> to issue a decree and order, to be effective immediately upon entry, declaring that each and all of the schools here involved are no longer to operate or function as a system in which some of the schools are in fact for white and some for colored students, and requiring that the schools begin immediately to operate and function as one unitary system of schools in which no person is effectively barred from any one of them because of his color.

Marshall Draft

Had the Black draft been issued as a dissenting opinion, its effect would have been most unfortunate. It would have broken what Justice White was to term "the semi-tradition of these school cases"[18]—that is, that of no dissent on the merits by any of the Justices. More than that, the tradition would have been broken by the Court's senior Justice, one of the giants in American legal history, who had been one of the staunchest supporters of desegregation. The Black dissent would have made the Court appear to be supporting Mississippi's resistance to *Brown* and to the commitment to racial equality.

However, it is probable that the Justice never intended his dissent to be issued. Instead, the circulation of his draft was a tactical device designed to ensure a decree in accordance with his strong views on immediate desegregation. As already indicated, Black knew that the others would be inclined to agree to such a decree if it were clear that the choice was to have him issue his draft as a dissent.

One would have thought that Justice Marshall, not only the sole black on the Court but the nation's most vigorous advocate of desegregation when he had headed the NAACP Legal Defense Fund, would be as firm as Justice Black in his opposition to the Chief Justice's draft order, which might have encouraged further dilatory tactics. Surprisingly, however, Marshall tried to work out a compromise which would have retained the essential elements of the Burger draft and still maintained the unanimity of the Court.

The Marshall draft was circulated early Monday morning, October 27, just before the next conference on the case, which had been scheduled at 10 A.M. that morning. The draft was headed "Memorandum to the Members of the Conference." It began: "Here are my suggestions for changes in the Proposed Order by the Chief Justice." The typed two-page Marshall order did not change the preamble of the Burger draft. As Marshall wrote, "these changes are suggested to replace the Order itself as contrasted to the preliminary paragraphs."

The copy of the Marshall draft in one of the Justice's files has the written note at its top: "attempt at compromise."[19] In its essentials, nevertheless, the Marshall order followed the Burger draft. Like the latter, it referred to the relief as "interim" and ordered the court of appeals to adopt plans to "provide reasonable means for achieving . . . immediate termination of a dual school system." The Marshall draft also contemplated delay. Under it, "the Court of Appeals shall enter its order on or before November 10, 1969, requiring the termination of the dual school systems and the establishment of unitary school systems on or before December 31, 1969."

Before circulating his draft order, Marshall had tried unsuccessfully to discuss the matter with Justice Brennan. Had he done so, it is likely that Brennan would have attempted to dissuade him from circulating the draft. After he read the draft, Brennan told Marshall that he regretted its circulation because it tended to undermine the position of those pressing for immediate and total desegregation. Marshall's law clerks told Brennan's clerks that the draft was intended as a compromise, since it permitted some delay but set time limits on its duration. As already noted, however, in its main points the Marshall draft followed the Burger proposed order and the ultimate time limit fixed by Marshall permitted delay in desegregation until the end of the year. By allowing that some delay was tolerable, such an approach risked future legal haggling over how much delay was proper.

Second Conference and Revised Draft Order

The Justices met for their second conference on the *Alexander* case on Monday morning, October 27, after circulation of the Marshall draft. At this point three draft orders had been circulated: those of the Chief Justice and Justices Black and Marshall. At the conference, the Chief Justice presented a revised draft of his proposed order. Copies of the revised Burger draft, on three typed pages, were passed out to the Justices when the conference began. The new draft contained some important changes. First of all, the proposed order began with a new paragraph: "1. The continued operation of segregated dual school systems being no longer constitutionally permissible under *Brown* v. *Board of Education*, 347 U.S. 483, the obligation of every school system is to achieve desegregation of such systems forthwith." Though there was no specific repudiation of the "all deliberate speed" formula (as there had been in the Brennan-Black draft), the Burger order did now expressly affirm the duty to desegregate "forthwith."

The revised draft also directed the court of appeals "to issue its decree and order, effective immediately, declaring that each of the schools here involved may not operate as part of a dual school system based on race or color, and directing that the schools begin immediately to operate as a unitary school system from which no person is to be barred because of race or color." The new Burger draft no longer contained any date by which the court of appeals should issue its order. Instead, the court was to act "effective immediately." This was a substantial step in the direction of the Brennan-Black draft.

There were also other changes in language in the Chief Justice's second draft designed to meet Justice Black's objections. There was no longer any reference to "interim" relief, and the phrase "at the earliest possible time" had been deleted; instead, as seen, the court of appeals order was to be "effective immediately." Further, that court was to accept the HEW plans and any modifications only if they insured "a totally unitary school system."

But the great defect of the new Burger draft to those who thought that the Chief Justice's first draft was not strong enough was the failure to change the preamble. At the conference Justice Black gave voice to the thought that must have weighed on the minds of all the Justices, that the case presented a direct conflict between the position of the Nixon administration and that of the Court. Black, and Justices Douglas and Brennan as well, said that the preamble was unacceptable because of its attempt to paper over this clash and because of the lack of a clear repudiation of the "all deliberate speed formula." The conference adjourned on the understanding that Chief Justice Burger would circulate another draft that afternoon. The understanding was that the Chief Justice would circulate simply a revised version of the draft order he had presented at conference. There would be an express rejection of the phrase "all deliberate speed" and the preamble would otherwise be made acceptable to those who agreed with Justice Black.

Burger Draft Opinion

The Justices expected the Burger third draft by 2 P.M. that afternoon. The new draft did not circulate until after 4 P.M. Both its form and content came as a surprise to the others. It will be recalled that the Chief Justice's covering memorandum transmitting his first draft order had stated that it was "to be followed by an opinion." In his October 26 memorandum, Justice Black had opposed this suggestion, asserting that no opinion was necessary. The matter was taken up at the October 27 conference, where (in the words of a written note in Justice Harlan's papers) the Court was "of view that no opinion should follow."[20]

Despite this, the third Burger draft order was accompanied by an opinion. "Following lunch today," stated the Chief Justice's covering *Memorandum to the Conference*, "I put my hand to a brief opinion and an order." The memo explained that though the Chief Justice "followed the precept that 'a committee cannot draft' orders and opinions efficiently, I was aided on the opinion by a rough draft which Justice White had put together for himself and by Justice Marshall's proposed order which was very much like what I had initially submitted." This last statement indicates that the Chief Justice himself believed that the Marshall draft was essentially similar to the first Burger draft order.

The Burger memo thus could state that "the order now proposed tracks both my early version and Justice Marshall's," except that the new Burger draft omitted Marshall's deadline dates. The Chief Justice wrote, "I do this, notwithstanding my belief that the order would be stronger with fixed dates. Open end directives have not worked too well since 1955."

This statement was inconsistent with what the Chief Justice had written in his covering memo to his first draft, where he had stated that he feared that inclusion of dates might "invite dilatory tactics." Nevertheless, the Chief Justice concluded his October 27 memo by asserting, "I believe the proposed opinion and the preamble to the order express every view of 'here and now' which anyone has proposed."

As already stated, the Burger October 27 memo was accompanied by a draft opinion and a revised draft order. The draft opinion, consisting of six and one-half typed pages, did not begin in the traditional manner with the name of the Chief Justice who had written it, but as an opinion issued in the name of each of the Justices. Hence, the draft opinion was headed:[21]

Opinion of the court by THE CHIEF JUSTICE, MR. JUSTICE
MR. JUSTICE   MR. JUSTICE
MR. JUSTICE   MR. JUSTICE   MR. JUSTICE
AND MR. JUSTICE

What the Chief Justice intended here is shown by the explanation in a *Memorandum to the Conference* sent around by him the next day, October 28: "If all agree, I suggest that we consider a 'Cooper and Allen' [*sic*], reciting of all members of the Court . . . because of the importance of the problem." The Chief Justice was proposing to follow the example set in the 1958 case of *Cooper v. Aaron*,[22] where the Court, for the first and only time, issued a joint opinion—a gesture that strikingly reinforced the Justices' unanimity in rebuffing what Chief Justice Warren was to term "Governor Faubus' obstructive conduct in the case."[23]

Despite its dramatic heading, the Burger proposed opinion was not impressive, even when considered only as a rough draft. Over five pages were devoted to a statement of the case. There were only four substantive paragraphs containing the decision and reasoning on the last page and a half of the draft. The first stated that the court of appeals "could fashion an order making effective prior to September 1, 1969, a plan of desegregation in each of the school districts here involved, at least as an interim measure. At this stage, substantially after *Green v. New Kent County, supra,* there was no occasion for further postponing the enjoyment of the constitutional rights of the school children in these districts."

The draft then declared, "The Court today enters the attached orders to give effect now to the long postponed rights of the pupils of these schools." But then Burger drew back, noting, "In the circumstances we have no doubt that this will present problems and difficulties." The difficulties were those faced by the draftsmen of the "interim plans" drawn by HEW, which "were prepared under difficult time pressures and, as the drafters conceded, were drawn without all the information needed in some situations."[24]

Yet, the draft went on, "If these plans are not administratively and academically perfect, they may be constitutionally tolerable." The court of appeals, examining them "as an interim relief measure . . . , may conclude that modifications are called for."

The Burger draft opinion then concluded with the following paragraph:

> The difficulties and defects such as may be encountered, including heavy burdens on pupils and teachers alike will, we hope, be more than offset by the fulfillment now to some of these pupils of promises long unkept. It is the Court's firm purpose to make clear that the language "all deliberate speed" fifteen years ago meant reasonable time to begin[25] not unreasonable time to finish, the task of providing bona fide unitary schools, not simply in these districts but in every part of every state.

To this reader the Burger draft was anything but a well-reasoned, tightly knit opinion; it read more like the rambling first draft by a new law clerk than the finished product of an experienced judge. It had no focus and, with the paragraph just quoted as its conclusion, it seemed to end in midair. But it was more than the style of the draft opinion that troubled Justices Brennan and Black, who had been pushing for a stronger order in the case. On Justice Brennan's copy of the Burger covering memo transmitting his draft opinion there is a handwritten note: "WJB had anticipated circulation of short form resolving where 'end of all delib speed' would go) See previous at-conf circulation by CJ. WJB thought there would be simply a revised version of that. Instead, this circulation persuaded WJB that the CJ was trying to save Nixon."[26]

Though this note is not in the Justice's handwriting, it appears to give an accurate picture of Brennan's reaction to the Chief Justice's draft opinion. In Brennan's view, I have been told, the draft opinion had the flavor of the Chief Justice's determination to avoid a confrontation with the Nixon administration. Instead of stressing the Court's determination to end dual school systems immediately, the draft pointed to the "problems and difficulties" faced by HEW and the court of appeals.

Third Draft Order and Brennan Redraft

As the Chief Justice's covering memo of October 27 pointed out, it was accompanied not only by the draft opinion, but also by a third draft of the proposed *Alexander* order. The new draft order contained the express statement in its preamble that, "it being clear that the standard of 'all deliberate speed' is not now a constitutionally valid basis for a dual school system based on race, the Court reaffirms the obligation of every school system to establish unitary school systems now." The preamble also changed the statement on the attorney general's urging immediate termination of dual school systems to "the Solicitor General having urged that the Respondent's obligation to desegregate their school systems is immediate and unqualified, urged implementation of a unitary system without awaiting the beginning of the 1970–71 school year"—a change which still indicated that the government had urged that no delay in desegregation should be permitted.

In other aspects, the third draft order was a backward step. Reconsideration by the Chief Justice of the Marshall "compromise" draft order had persuaded him that he could attempt to steer the *Alexander* order back to his own first draft in some crucial respects. Thus, the third Burger draft again contained references to "interim relief" to be ordered by the court of appeals as well as for that court to "provide reasonable means to achieve termination of a dual school system forthwith." In addition, despite the disclaimer in the Burger covering memo, the new draft order did contain a cutoff date for execution of the court of appeals order. Though there was no deadline specified for entry of orders by the court of appeals, the new Burger draft order did provide that that court was to enter "such orders or decrees, pending final resolution of this litigation, declaring that public schools may not operate as a dual school system based on race and directing that such schools begin to operate as a unitary system no later than November ____, 1969."

Nothing further was done on Monday, October 27, after the Burger draft opinion and third draft order were circulated after 4 P.M. that day. The Chief Justice did, however, remind the Justices that the preliminary meeting of the judicial conference would begin Thursday morning, and that the conference would last a week. If, as all had agreed at the first conference, a speedy decision was essential, it would have to be issued no later than Wednesday. The alternative was that it be delayed for at least a week. This time schedule strengthened the pressures for an early agreement in the case.

The strongest opposition to the Burger drafts came from Justices Black and Brennan. They both feared that the Burger draft opinion all but nullified the statement on "all deliberate speed" that was now contained in the third draft order. "For me," Brennan wrote to the Chief Justice on the morning of October 28, "the prime objective of what we file in these cases is to remove the impression of HEW and the Justice Department that the standard of 'all deliberate speed' retains some vitality. I fear that the message is obscured by your proposed opinion."

Black and Brennan also strongly objected to the revival of the references to "interim relief" in the revised draft order. Their objection was based on the phrase's intimation that any desegregation so ordered would be provisional and that the right to final relief had not fully matured. The refusal to characterize the immediate relief as "interim" did not mean that the two Justices were opposed to modifying it once it was in operation. The relief was "final" because the petitioners had a right to final relief immediately, and the court of appeals could draw on existing HEW proposals to frame effective realistic desegregation plans. Once in effect, the relief would be provisional since it could be modified if necessary; but that is true of any equitable relief.

Justice Black also opposed any specific dates in the *Alexander* order. He did so for two principal reasons. The first was that provision of a period for compliance, even with a terminal date, invited delay because on remand Mississippi could argue that the Court did not have all the facts before it and that on all the facts a further extension of time for compliance was warranted. In addition, any time period sufficiently short to satisfy all the Justices might be too short to permit compliance, and thus the Court would "look silly." The Court might also "look silly" if the specified time for compliance passed without compliance. It should be noted that, on similar grounds, Justice Black objected to any reference to desegregation "plans." Accordingly, the word was carefully avoided in most of the drafts that were written and does not appear in the *Alexander* per curiam as finally issued.

Early on Tuesday morning, October 28, Justice Brennan prepared a new version of the second Burger draft order, the draft which the Chief Justice had distributed at the Monday conference. The new Brennan draft was circulated about 10 A.M., in a letter that the Justice sent to the Chief Justice, with copies to the others. This is the letter that has been quoted on the "prime objective" of the Court's decision in Brennan's view—to get across "the message" that the "all deliberate speed" standard no longer retains any vitality at all. "My view," Brennan wrote, "is that we should state the message in the briefest and plainest possible words. The proposal you circulated at Conference yesterday based on Hugo's suggestions strikes me as a model upon which to build." Brennan did not, of course, tell the Chief Justice that the original draft of the order which contained "Hugo's suggestions" had been telephoned by him to Justice Black's secretary.

The Brennan letter concluded with a suggested draft order, which began with the heading "Per Curiam" (thus rejecting the Burger idea that the *Alexander* decision should follow the precedent of the 1958 Little Rock case, *Cooper* v. *Aaron,* and be issued in the name of each of the Justices, listed seriatim).

The new Brennan draft order began with an entirely new preamble, intended to eliminate what both Brennan and Black considered the major defect of the Burger draft orders. The Brennan preamble declared flatly that "desegregation of segregated dual school systems according to the standard of 'all deliberate speed' is no longer constitutionally permissible. The obligation of every dual school system is to desegregate now." There was no reference to the position of the government. Instead, the Brennan preamble ended by holding that "the Court of Appeals should have denied the motion and directed that each school system begin immediately to operate as a unitary school system within which no person is to be barred from any school because of race or color."[27]

Brennan's proposed order did not contain any deadlines or other dates. Compliance was to be "immediate" and final. There was no reference to "interim" relief. The text of the order consisted of five paragraphs, which were substantially similar to those contained in the final *Alexander* order. They provided for the court of appeals to issue an order, effective immediately, directing the operation of unitary school systems within which no person was to be barred because of race or color. The HEW recommendations might be accepted, with modifications to "insure a totally unitary school system." The district court was barred from interfering in the relief; the court of appeals was to retain jurisdiction to ensure compliance with its order; and the court of appeals delaying order was vacated and that court was to lay aside all other business to carry out the Supreme Court's mandate.

Final Draft

After Justice Brennan had circulated his proposed order on Tuesday morning, October 28, he telephoned Justice Douglas in Elmira, New York, where Douglas had a speaking engagement. Douglas agreed to join the Brennan draft. Later that morning Justice Black sent a letter to the Chief Justice stating that he was unable to agree with the Burger opinion and order. "I do substantially agree, however to the suggested Per Curiam of Brother Brennan circulated this morning."

Shortly after Black's letter was sent, Justice Harlan circulated a letter to the Chief Justice which said that "the most satisfactory disposition of the case" would be that in the proposed order which had been drafted by Justice Marshall, "but unaccompanied by an opinion as suggested in your second circulation of yesterday." This was the letter in which Harlan indicated his disagreement, already quoted, with the Burger preamble because it tended to "blink the fact that the Government stands in opposition to the central and only issue in the case."

Harlan noted that the dictation of his letter was under way when the latest Brennan proposal was received. He wrote that the difference between his and the Brennan view was that he preferred "the explicit provision of 'outside' dates" (because "such dates will strengthen the hand of the Court of Appeals in resisting dilatory tactics") and the use of the words "interim" and "terminal." However, Harlan wrote, "If these factors continue to stand as road blocks to obtaining as much unanimity among us as possible, then I would be prepared to cast my vote for the Brennan proposals." As it turned out, this was the most important sentence in Harlan's letter, for it ultimately meant another vote for the Brennan draft.

The next development was a draft opinion which was circulated by Justice Stewart on Tuesday afternoon. The Stewart draft was an opinion on nine typed legal-size pages, with no separate order. As Stewart's covering letter explained it, "Instead of setting out a separate order at the end of the Per Curiam, I have written the terms of our mandate into the text of the proposed opinion, in accord with our conventional practice."

The Stewart draft contained a lengthy history of the case. It declared that "further delay, no matter from what source derived, will not be tolerated"—though it did provide a cutoff date for the final decree ("in no event later than November 15, 1969"), as well as up to ten days following the decree to enable the parties to comply. Stewart also expressly rejected "all deliberate speed," despite the statement in his letter, "I find difficulty in perceiving how the phrase 'all deliberate speed' really has much to do with these cases." As Stewart explained it, "In deference to Hugo, Bill Brennan, and perhaps others, I have included in the enclosed their language about 'all deliberate speed.' "

What Justice White termed "the flurry of paper"[28] continued with letters to the Chief Justice from Justices Harlan and Marshall. Harlan now wrote that he was also prepared to join the Stewart opinion "should it commend itself to a majority." Marshall wrote that, despite his own preference, "I now assume that it is impossible to get unanimity on cutoff dates. On that assumption I could agree to the draft of WJB."

With the Harlan and Marshall letters Chief Justice Burger realized that he could not get a Court for his draft opinion. No one had indicated a willingness to join and it had bitterly offended some of the Justices. His latest draft order was also most unlikely to command a majority, much less the unanimous Court that had become the tradition in desegregation cases. The Chief Justice also knew that the Brennan draft order was now acceptable to a majority. It had the firm support of Justices Brennan, Black, and Douglas and the acquiescence of Justices Harlan and Marshall. The adherence of the latter two was particularly significant, since they had, from the Friday conference until then, favored some period for compliance.

At this point, at about 4 P.M. on Tuesday, October 28, the Chief Justice went to the Brennan chambers and conferred with the Justice. They went over the Brennan draft together and made some unimportant changes. Brennan and Burger then agreed on a revised version of the Brennan draft, which did not depart in any substantial respect from the draft order sent around by Brennan that morning. The Chief Justice then circulated the revised version of the Brennan draft upon which he and the Justice had agreed. This was very similar to the final order issued in the case.

With the revised draft went a covering *Memorandum to the Conference:* "Enclosed is 'another try' in light of various proposals received. It returns to what I proposed to the Conference except (a) the preamble is altered and (b) the dates are omitted." As trivial as these changes were made to seem, they were crucial and meant an order like that circulated by Justice Brennan. This was noted on Justice Harlan's copy of the Burger memo, where there is a handwritten note: "almost identical to WJB draft of Tues. AM."

A last draft was circulated the next morning, Wednesday, October 29. The Chief Justice had received a number of suggestions and had incorporated some of them in the final draft. All were sent to Justice Brennan for his consideration before the final order was drafted.

In his covering *Memorandum for the Conference,* the Chief Justice wrote that the final draft "resembles the proverbial 'horse put together by a committee' with a camel as the end result. But then even the camel has proven to be useful." The memo also recognized that there had been no support for following the *Cooper v. Aaron*[29] precedent. Hence, the Burger memo stated, "There is some view, which I now tend to share, that a recital of all names at the head of the order has a tendency to give it undue emphasis. I will therefore have this entered as a routine order and decree letting the contents convey their own urgent message."

Even before the final draft was circulated, Justice Harlan wrote to the Chief Justice that he was prepared to concur in the revised Burger-Brennan draft. His letter, sent late in the afternoon of Tuesday, October 28, stated, "We have reached the point in our deliberations where the differences amongst us hang not on any matters of substance but on pure semantics. Frankly, I think the important thing now is to reach an agreement on *some* disposition which can be announced at the earliest possible moment, preferably not later than tomorrow afternoon."

After the final draft was circulated, the Justices all quickly agreed to join. Justice Brennan went to the Douglas chambers with the draft. Douglas wrote "I agree" on his copy and gave it to Brennan. Later that morning, at 11:30 A.M., Brennan telephoned Justice Black at home and read him the final draft. Black approved it and Brennan then went to the Chief Justice and told him that Black, Douglas, and Brennan agreed to join. At 12:55 P.M., Justice Marshall came to see Brennan and said that he also agreed. Marshall then sent a letter joining the draft to the Chief Justice. Justices Harlan, White, and Stewart also wrote to join the order.

In their letters of agreement, White and Stewart expressed what Stewart called "some substantial misgivings."[30] Nevertheless, White wrote, "In the semi-tradition of these school cases, which any Justice may disown if he chooses, I shall not dissent from this order." White wrote that he did this despite deficiencies in the order which he noted, particularly the failure to specify when the court of appeals should issue its order.

In White's view, what the Court was saying was "that the deliberate speed formula has been abandoned (which we have said before) and that as soon as possible is an adequate substitute." White then wrote, "Hugo is convinced that a mistake was made in 1954–55 with respect to the deliberate speed formula. I am beginning to understand how mistakes like that happen. Nevertheless, I join, expecting the Court of Appeals to make sure that the shortcomings of this order never come to light, even though it is that Court which entered the order which we now find unacceptable."[31]

Justice White's tepid letter of agreement did not, however, end the maneuvering over the *Alexander* order. Though, Justice Black had given his agreement to Justice Brennan over the phone, after talking with his law clerks later in the day, he became disturbed over the substitution in the final draft of a request that the court of appeals "give priority to this mandate" for a request that it "lay aside all other business of the Court to carry out this mandate immediately." He felt that the change might lead the court of appeals to delay. Justice Brennan discussed the matter with him and urged that the past history of the fifth circuit court of appeals showed that it was fully entitled to the Supreme Court's confidence in desegregation matters. Moreover, Brennan said the priority language was more in keeping with another change that had been made with Black's approval. Under it, the court of appeals, instead of being "directed" to act, was now "requested" to do so.

Justice Black finally became reconciled to the change when the phrase "the execution of" was added so that the court of appeals was enjoined to "give priority to the execution of this mandate." The word "mandate" itself became "judgment" in the final order, a stylistic change made by the Court's deputy clerk.

The Court could now finally issue its *Alexander* order. It was announced, in substantially the form drafted by Justice Brennan, as a unanimous per curiam in the early evening of Wednesday July 29, less than a week after the argument in the case.

"Yes, Virginia, There Is a Constitution"

The *Alexander* decision, with its specific repudiation of the "all deliberate speed" standard, its reversal of the court of appeals August 28 order, and its decree ordering that court to direct the school districts to "begin immediately to operate as unitary school systems," was widely viewed as a direct rebuff to the Nixon administration, the very result Chief Justice Burger had tried so hard to avoid in his drafts in the case. The Chief Justice had taken the case for himself, although he had not voted with the majority. The Justices had, however, refused to accept the Burger drafts and the Chief Justice himself had had to accept the strong order prepared by Justice Brennan, which flatly rejected the new administration's position on delay.

Well might the *St. Louis Post-Dispatch* headline its editorial approving the decision "Rebuke by the Supreme Court,"[32] a theme repeated in news accounts throughout the country. Thus, *Newsweek* asserted, under the heading "Yes, Virginia, There Is a Constitution," that the decision was "a stinging rebuke to the go-slow tactics of the Nixon administration, which had just installed Warren Burger at the helm of the high tribunal."[33]

The *Newsweek* article also stated, "Some veteran Court watchers thought they saw signs of the hand and mind of Justice Hugo Black in the wording of the order." The same point was made in newspaper accounts.[34] Apparently someone in the Court who had compared the *Alexander* order with the order proposed in Justice Black's draft dissent had informed the press that the Justice had played a major role in the drafting process. Only Justices Black and Brennan knew that the order in the Black draft dissent had originally been written by Brennan himself.

5

Swann before the Supreme Court

To the press and the public, *Alexander v. Holmes County Board of Education*[1] was a clear signal that the Burger Court was as committed to school desegregation as the Warren Court had been. To the *Green*[2] requirement of a school plan "for prompt and effective disestablishment of a dual system . . . that promises realistically to work, and promises realistically to work *now*,"[3] was added the *Alexander* repudiation of any further delays, even though they were requested by the government itself because "time was too short and the administrative problems too difficult to accomplish a complete and orderly implementation of the desegregation plans."[4] Alexander required orders "directing that they begin *immediately* to operate as unitary school systems." All motions for additional time must be denied because the "all deliberate speed" standard "is no longer constitutionally permissible." The duty " of every school district is to terminate dual school systems *at once* and to operate now and hereafter only unitary schools."[5]

To those familiar with the *Alexander* decision process, on the other hand, there were signs that the new Chief Justice would not play the same role in school desegregation cases as had his predecessor. Chief Justice Warren had led the Justices in their efforts to abolish racial segregation—from the *Brown* case, where he had forged the united Court that finally struck down segregation, to the *Green*[6] and *Montgomery County*[7] cases, where the desegregation efforts of the Warren Court had culminated. If, as Warren wrote to Justice Brennan, with the former case "the traffic light will have changed from *Brown* to *Green*,"[8] that was true, in large part, because of the Chief Justice's own leadership.

Alexander indicated that the situation would be different under Warren's successor. Far from attempting to lead the Court to its categorical rebuff of the Nixon administration, the new Chief Justice had attempted to mold the Court's decision (again using Justice Harlan's phrase) "to blink the fact that the Government stands in opposition."[9] And he had sought to weaken the *Alexander* order itself by providing only for "interim relief" to be made "operative" only "at the earliest possible time."[10] The Burger efforts had been frustrated by the united opposition of most of the Justices, including the normally conservative Harlan. *Alexander* showed clearly that the new Chief Justice was no Earl Warren so far as the Court's role in desegregation cases was concerned.

Northcross v. Board of Education

The different position of the new Chief Justice was again demonstrated a few months later in the Memphis school case, *Northcross v. Board of Education*.[11] In May 1969 the district court had ordered the Memphis school board to submit a new desegregation plan by January 1, 1970, based on geographic pupil assignments and transfer provisions. Petitioners appealed to the court of appeals for a stronger plan. After the Supreme Court's *Alexander* decision, petitioners moved for an injunction directing the district court "to prepare and file on or before January 5, 1970, in addition to the adjusted zone lines it is presently required to file, a plan for the operation of the City of Memphis public schools as a unitary system during the current 1969–70 school year." The court of appeals denied the motion on the ground that *Alexander* was inapplicable because the fact pattern there "is not descriptive of the present situation of Memphis." Instead, the court was "satisfied that the respondent Board of Education of Memphis is not now operating a 'dual school system' and has, subject to complying with the present commands of the District Judge, converted its pre-*Brown* dual system into a unitary system 'within which no person is to be effectively excluded because of race or color.' "[12]

On January 30, 1970, petitioners filed a petition for certiorari. The case was discussed by the Justices at several conferences. The key conference was held on Friday, February 27. As at previous conferences, there was disagreement about how to handle the case. Chief Justice Burger spoke in favor of granting certiorari and setting the case for argument. The Court could then reach the issue of what was required for a school system to be unitary. The Chief Justice was supported by Justices Harlan, Stewart, and White. The other three Justices—Black, Douglas, and Brennan (Marshall did not participate because he had been involved in the Memphis situation when he had been solicitor general)—disagreed. Their position was led by Justice Brennan, who urged the conference to avoid reaching the issue because the Court could not give a comprehensive definition of a "unitary school system" that would be satisfactory in different types of cases. There was always the risk, Brennan argued, that any "realistic" definition by the Court would appear to be a retreat from *Brown* and any other type of definition would, given the views of most whites, simply be impractical.

Even though there appeared to be four votes to grant certiorari and reach the definition issue, the conference was so sharply split that a committee of Justices Brennan, Stewart, and White was appointed, in the hope that they could reach an agreement on the case. Despite his minority position at the conference, Justice Brennan drafted a per curiam opinion over the weekend. He sent copies to Justices Stewart and White with a "Dear Potter and Byron" letter: "I thought our conference might move along better if we had something on paper to discuss. Accordingly I enclose a proposed *per curiam* which reflects my thoughts about a disposition."[13]

The Brennan draft was essentially the per curiam that was issued in the *Northcross* case. Certiorari was granted and the findings of the district court that the dual school system had not been dismantled and that the school board's plan did not have real prospects for dismantling it were upheld as supported by substantial evidence. Hence, the court of appeals exceeded its appellate power in substituting its own finding that the board was not operating a dual system. The court of appeals "erred in holding that *Alexander* . . . is inapplicable to this case." The case was remanded "with direction that the District Court proceed promptly to decide the case consistently with *Alexander*."[14]

By deciding that the court of appeals erred in reversing a district court finding that was supported by substantial evidence, the Brennan draft was able to avoid the issue of whether the Memphis board was, in fact, operating a dual or unitary school system. This enabled it also to avoid the issue of what constituted a unitary school system.

Justices Stewart and White met with Justice Brennan in his chambers after the Court session on Monday, March 2. Despite their support of the Chief Justice at the conference, Stewart and White agreed to a revised version of the Brennan draft, which differed from the original only in minor details. This draft was circulated two days later.[15] Later that day, the per curiam was joined by Justices Black and Douglas. This gave it a majority, since only seven Justices sat in the case (there having been no replacement yet for Justice Fortas, who had resigned, and Justice Marshall having, as already noted, disqualified himself). The next day, March 5, Justice Harlan made a sixth for the Brennan draft.

A day later, however, the Chief Justice circulated an opinion "concurring in the result." It was accompanied by a *Memorandum to the Con-*

ference, which stated, "Just before departing on Thursday I put together the thought expressed in the attached draft. I think it is desirable for several reasons, not all of which are related to the particular case."

The Burger draft was essentially the same as the concurrence that he issued in the case. It reiterated the view that the Chief Justice had expressed at the conference. "Save for one factor, I would grant the writ and set the case for argument." The factor that led him to concur "in this particular case is that Mr. Justice Marshall would not be able to participate." Aside from that, Burger wanted full argument and review because "the time has come to clear up what seems to be some confusion, genuine or simulated, concerning this Court's prior mandates."

On the definition issue, the Chief Justice asserted, "The suggestion that the Court has not defined a unitary school system is not supportable." He then referred to the *Alexander* statement "that a unitary system was one 'within which no person is to be effectively ex[c]luded from any school because of race or color.'" Burger did, however, concede that this was stated "perhaps too cryptically." Despite *Alexander* and other cases, the Court had left open the detailed requirements that would make a school system unitary.

According to the Chief Justice, this was a lacuna which the Justices should fill. "At some point we should resolve some of the basic pracical problems including whether any particular racial balance must be achieved as a constitutional matter, to what extent school districts and zones may or must be altered and to what extent busing is compelled as a constitutional requirement."[16]

These were precisely the questions that, as we shall see, the Chief Justice was to attempt to answer in his draft opinion of the Court in the *Swann* case. They were not dealt with at all in *Northcross* because the Brennan draft per curiam was joined by the other Justices. The Chief Justice circulated two additional, slightly revised versions of his separate opinion. When *Northcross* was decided on Monday, March 9, Burger alone did not concur in the per curiam. His opinion concurring in the result, raising the questions quoted in the preceding paragraph, suggested that the Court was not as united as the 7–0 result in *Northcross* indicated. This intimation was borne out by the decision process in *Swann.* That case was then before the court of appeals and was soon to make its way to the Supreme Court.

Certiorari and Scheduling *Swann*

On May 26, 1970, shortly after *Northcross v. Board of Education* was decided by the Supreme Court, the U.S. Court of Appeals for the Fourth

Circuit handed down its decision in the *Swann* case.[17] As seen in Chapter 1, the court affirmed Judge McMillan's order for secondary schools, but vacated his order respecting elementary schools. The case was remanded to the district court for reconsideration and submission of further plans. A few weeks later, on June 18, the *Swann* plaintiffs filed a petition for certiorari. The Justices voted unanimously in favor of the writ, and it was granted on June 29.[18] In its order granting certiorari, the Court directed reinstatement of Judge McMillan's order pending further proceedings in the case.

On remand, the district court received two new plans for the elementary schools, one from the U.S. Department of Health, Education and Welfare and one prepared by four members of the nine-member Charlotte-Mecklenburg Board of Education. After a lengthy hearing Judge McMillan directed the board to adopt a plan or the court-ordered plan prepared by Dr. Finger (the court-appointed expert) would remain in effect. On August 7, the board indicated it would "acquiesce" in the Finger plan, though it reiterated its view that the plan was unreasonable. The same day, Judge McMillan issued an order directing that the Finger plan remain in effect.[19] This meant that the McMillan order of February 5, 1970, with its provision for busing of schoolchildren at all levels, would go into effect when the new school term began in September.

The imminence of extensive busing in Charlotte led to substantial attempts to influence the Court's decision process. Even before the *Swann* case reached the high bench, President Nixon issued a lengthy statement in which he urged that children should be permitted to attend their neighborhood schools. The decisions by lower courts ordering busing "of pupils beyond normal geographic school zones for the purpose of achieving racial balance" went far beyond what the Constitution required. "Unless affirmed by the Supreme Court," the president declared, "I will not consider them as precedents to guide administration policy elsewhere."[20]

In the Court itself, on the other hand, Justice Douglas had circulated a strong memorandum in favor of busing on June 27, just before certiorari had been granted. Headed "Memorandum from Mr. Justice Douglas," it contained seven printed pages and urged that busing "is a permissible tool for ending desegregation [*sic*]." As Douglas saw it, "The federal court should have power to enter a decree requiring busing. Otherwise the State gives the black no real choice but to stay in his *de jure* segregated school. That is denying access to the quality school by reason of race."

In the South, of course, there was a continuation of the outcries against the Court that had begun with the *Brown* case. In particular, southern politicians attacked the Justices for not interrupting their summer vacations to decide the crucial issue before the school term began. On August 7, G. Harrold Carswell, the Southerner on the U.S. Court of Appeals for the Fifth Circuit whose nomination to the Supreme Court had been rejected by the Senate a few months earlier, sent a telegram to each of the Justices. It made "a humble plea" to the Court "to halt artificial integration and to stop social experimentation with our school children." Carswell called for early consideration of the *Swann* case: "I respectfully urge the Supreme Court members to sacrifice a portion of their vacation to convene in extraordinary session to meet this responsibility on behalf of education in our nation."

Toward the end of August, the Charlotte school board requested a stay of Judge McMillan's August 7 order directing that the Finger plan go into effect. The Chief Justice telephoned the Justices who were in Washington and sent telegrams to the Justices who were out of town, requesting their views on a stay. None of them voted in favor and the stay was accordingly denied.

Several days later, the Justices dealt with the desire to have the case decided as early as possible. The Chief Justice sent telegrams on August 26: "Please send me your vote re expediting consideration." Burger recommended that *Swann* "should be heard on October 12, the first day of oral argument if possible." Justice Black replied by letter the same day that he favored setting the case for argument "at the earliest possible moment (even before October 12th and favor whatever order is necessary to provide for that early hearing)."

Justice Harlan replied on August 27, saying that he agreed with the Chief Justice. "I see no useful purpose in trying to set the cases for an earlier date than October 12, as I gather Brother Black suggests might be done. Indeed, I think it would be a mistake to do so in view of the already short briefing period involved." The others agreed with Burger and Harlan, including newly appointed Justice Blackmun, who wrote from Minnesota on his eighth circuit court of appeals stationery.

There was a peculiar "Dear Chief" letter from Justice Douglas. Sent September 4, it began: "For some inexplicable reason your telegram of August twenty-sixth did not reach me at Goose Prairie until last evening when it arrived in a mail bag full of certs. Hence my delay in replying to your questions. I apologize."

Douglas wrote, "My preference is to leave all of these cases, including No. 281—*Swann* v. *Charlotte-Mec[k]lenburg*, to the Court of Appeals. My review of the Courts of Appeals decisions in these school cases this summer have impressed me and I have concluded they are doing a very responsible job. So my preference would be to hear none of them, not even No. 281—the *Swann* case." The other cases referred to were other school desegregation cases which were then pending in the Supreme Court. Douglas also stated, "I have done quite a lot of work on the *Swann* case and I do not see how we could possibly do anything more than affirm when we reach the merits."

On September 9, the Chief Justice circulated the following *Memorandum to the Conference—Re: No. 281—Swann v. Charlotte-Mecklenburg*: "Supplementing Mr. Justice Douglas' memorandum to the Conference of September 4, the following was added: 'I would deny cert.' "

It is difficult to know what to make of the Douglas letter, as supplemented by the Burger memo. Douglas's statements that he did not want to hear the cases (including *Swann*) and would leave them to the courts of appeals and that the Court could only affirm in *Swann* were contrary to every view otherwise expressed by him in the case, from his already mentioned June 27 memo to the statements, memos, and drafts which, as will be seen, he was to issue during the *Swann* decision process. The supplementary statement that Douglas "would deny cert" is even more puzzling. Certiorari had already been granted in *Swann* and Douglas himself had voted in favor of the grant on June 29.

At any rate, in accordance with the Chief Justice's proposal, the *Swann* case was scheduled for oral argument on October 12, 1970, the first argument day of the new 1970 Term. The order scheduling *Swann* and the other cases to which Douglas had referred was reported by the Associated Press in a release published on September 1: "Apparently stung by critics who say the court is moving too slowly, the Justices set aside the first argument day of their 1970–71 term for the hearing and ordered the lawyers in the six cases to speed the filing of their briefs. Still, the court declined to interrupt its 13-week recess for a special session." A copy of the AP release was circulated by the Chief Justice the next day.[21]

Oral Argument

Like all Supreme Court sittings, the Monday, October 12, session began at 10 A.M. when the entrance of the black-robed Justices through the red velour draperies behind the bench was announced. At the sound of the gavel all in the crowded courtroom rose and remained standing while the Court crier intoned the time-honored cry, "Oyez! Oyez! Oyez! All persons having business before the Honorable, the Supreme Court of the United States, are admonished to draw near and give their attention, for the Court is now sitting. God save the United States and this Honorable Court."

The Court Chamber itself is the most impressive room in the building. It measures 82 by 91 feet and has a ceiling 44 feet high. Its 24 columns are of Sienna Old Convent marble from Liguria, Italy; its walls are of Ivory Vein marble from Alicante, Spain; and its floor borders are of Italian and African marble. Above the columns on the east and west walls are carved two marble panels depicting processions of historical lawgivers. Of the eighteen figures on the panels only one is famous as a judge, and he is the one American represented: John Marshall. His symbolic presence strikingly illustrates the Supreme Court's role as primary lawgiver in the American system.

The room is dominated by the Justices' long, raised bench, straight in 1970 (two years later, the Chief Justice had it altered to its present "winged," or half-hexagon, shape). Like all the furniture in the room, the bench is mahogany. Behind the bench are four of the massive marble columns. A large clock hangs on a chain between the two center ones. In front of the bench are seated, to the Court's right, the pages and clerk, and, to the Court's left, the marshal. Tables facing the bench are for counsel. Behind the tables is a section for members of the bar and a much larger general section for the public, with separate areas for the press and distinguished visitors.

Goose-quill pens are placed on counsel tables each day that the Justices sit, as was done at the earliest sessions of the Court. The practice had been interrupted by World War II, when the prewar supply ran out,[22] and then again in 1961, when the quills were temporarily replaced by more modern writing instruments.[23] But traditions die hard at the Supreme Court. The quills soon found their way back to the counsel tables, and there are still spittoons behind the bench for each Justice and pewter julep cups (now used for their drinking water).

On October 12, after the Justices took their seats in their plush black-leather chairs, the Court first announced orders in eight cases granting review and summarily disposed of twenty others—affirming, reversing, or vacating the lower court judgments, without full briefs and arguments. Then the Chief Justice leaned forward and said, in his mellow bass, "The first case on for argument this morning is No. 281, *Swann* against *Charlotte-Mecklenburg Board of Education.* . . . Is counsel ready? Mr. Chambers, you may proceed whenever you are ready."[24]

At this, Julius LeVonne Chambers stepped to the lectern and began with the traditional opening, "Mr. Chief Justice, and may it please the Court."

Oral arguments in the Supreme Court are often dramatic events, participated in by the leaders of the American bar. Chambers was certainly not one of these. He was still a largely unknown lawyer from North Carolina, 34 years of age, who was arguing his first Supreme Court case. But he had lived with the case from its beginning and had a more intimate knowledge of its details than almost anyone else. In addition, as a southern black who had excelled at the best law school in his state, he could serve as a striking example of what could be accomplished once racial barriers to education were removed.

Chambers started by summarizing the facts and history of the case. He had barely gotten under way when the Justices began to pepper him with questions. In fact, except for the first few minutes of his presentation, Chambers spent virtually all of his time responding to the constant queries directed at him from all sides of the bench. The transcript of the Chambers oral argument fills 24 printed pages. During it, the Justices interrupted Chambers 114 times; 103 of the interruptions contained questions, all in a space of less than an hour.

The incessant interruptions from the bench can make things most difficult for even the seasoned Supreme Court advocate—and even more so for a neophyte like Chambers. But that is now the normal scenario for arguments before the highest bench. The days of the great advocates of the past, when Daniel Webster or William Wirt would give virtuoso performances extending over several days, have long since been gone. Supreme Court arguments now are less solo presentations than Socratic dialogues in which bench and bar play an almost equal part.

What makes this so difficult for counsel is that each side's time is severely limited. In the Court's early years, there were no time restrictions and arguments ran as long as fourteen days. Those days, too, are long gone. The Court now allows each side only thirty minutes for ordinary cases on its docket. More important cases may be given additional time. In *Swann,* over three hours was allowed for the oral argument, an extension that still scarcely gave counsel an adequate opportunity to deal with so contentious and complicated a case, particularly since most of the time allowed had to be used to respond to the incessant questioning by the Justices.

Counsel is normally not allowed additional time because the interruptions from the bench have made it all but impossible to present more than a fraction of his prepared arguments. Regardless of how much has been covered, the time limits are strictly observed. When counsel has used up his time, a red light goes on at his lectern; he must then stop, whether he has finished or not; if he tries to continue, the Chief Justice will stop him in the middle of a sentence if need be. The *Swann* argument ended while James M. Nabrit III, the second attorney for petitioners, was still making his presentation. "I think your time is up, Mr. Nabrit," the Chief Justice interposed. "Thank you for your submission."

Nabrit could scarcely have been mollified by the Chief Justice's earlier interjection: "Let me interrupt you a minute to say that we have allowed the other side of the table about four minutes extra, and so we'll allow you to run until 3:00 o'clock, which will give you a little bit more time, in view of our balancing of time problem, anyway. [General laughter.]"

The difficulty for counsel is compounded by the fact that while the argument and questioning proceed, the bench itself is rarely still. The Justices, much to the consternation of a Court neophyte like Chambers, conduct often lengthy whispered conversations with their colleagues, thumb through the record and other documents, talk to their clerks, send pages on errands, write memos—even doze at times. Justice Douglas, in particular, was noted for not paying attention to what counsel was saying. During oral arguments, in fact, Douglas seemed to spend most of his time writing letters. You could tell he was writing letters because, every once in a while, he would lift his head from his scribbling and very ostentatiously lick an envelope from side to side and seal it. At the same time, every now and then he would straighten up and direct an acute question to counsel. As one of the Justices once put it, "Bill could listen with one ear."

Despite these difficulties, however, Chambers made some telling points in his October 12 presentation. He stressed that only the plan approved by Judge McMillan would do away with the all-black schools in Charlotte. "It would," he declared, "be a rejection of the faith that black children and parents have had in *Brown*—the hope of eventually obtaining a desegregated education—for this Court now to reverse the decision of the District Court, and to now adopt 16 years after *Brown* a test that would sanction the continued operation of racially segregated schools."

Chambers urged that the constitutional requirement meant that no child should attend a "racially identifiable" school—which he defined, in response to Justice Stewart, as a school with 90 percent white or over 50 percent black students. When asked to summarize the legal issue by Justice Harlan, Chambers said that it was whether a school board could continue to perpetuate segregated schools, when they were segregated by the government, if there was a feasible plan to desegregate in existence. The answer, he suggested, was that they could not.

Justice Black then interposed. "Is there a constitutional requirement to bus pupils and to force states to buy a large number of buses?" The answer was, "The Constitution allows it."

Mr. Justice Black: "Are you saying that under certain facts the Constitution requires busing?"

Mr. Chambers: "The Constitution requires whatever is necessary to desegregate."[25]

After Chambers had finished, Erwin N. Griswold, the solicitor general, appearing for the government, amicus curiae, spoke. Griswold was a brusque former Harvard Law School dean who had been appointed in an effort to add prestige to the Nixon Justice Department. He advocated a go-slow policy to desegregation. Judge McMillan had gone too far, he said. The objective was not to establish a racial balance, but to disestablish a dual school system. "The Equal Protection Clause and the Due Process Clause of the Fourteenth Amendment require the latter standard. These are the only constitutional principles involved, and this is not a retreat."[26]

Griswold also argued for less busing than Judge McMillan had ordered. Justice Harlan asked, "You do not contend, I take it, that busing, as such, is an impermissible remedial measure?"

"No, Mr. Justice," was the answer. "It becomes a question only with the amount, and the distance, of the busing." The government had no problem with respect to the secondary schools. "The problem arises exclusively with respect to elementary schools. . . . You're dealing with very small children," who should not be bused long distances "from their home areas." Later in his argument, he quoted President Nixon's already cited statement, emphasizing that the president had said neighborhood schools are the "most appropriate base" for disestablishing a dual school system.[27]

The remainder of the session, after the solicitor general had finished, saw the lawyers for the school board, William J. Waggoner and Benjamin S. Horack, argue that the board had done its constitutional duty. Asserted Waggoner, "The Board has gone further than I perceive the constitutional duty to be." Then Mr. Nabrit spoke in support of petitioners. At one point he stated that the "neighborhood school concept in the context of Charlotte is really a fiction. It never existed in Charlotte." At this, Justice Black asserted, "I think there is something to the concept of the neighborhood schools." Nabrit replied that the neighborhood school, whatever its merits, had no standing when measured against the constitutional rights it would violate.

Black came back with an emotional plea:

From the first case, I have been interested in plain discrimination on account of race. We should correct that. But it disturbs me to hear we should try to change the whole lives of people around the country. You're challenging the place people live. You want to haul people miles and miles and miles in order to get an equal ratio in schools. It's a pretty big job to assign to us, isn't it? How can you rearrange the whole country?

Nabrit answered that he was not asking that; he was just seeking integrated schools.[28]

The *Swann* argument ended at 3 P.M. Since there had been an hour recess for lunch at noon, it had taken over three hours.

The next step, after the argument, was for the Justices to consider the case at their weekly Friday conference. Before they could do so, each of them received an October 14 letter from Georgia Governor Lester Maddox. The letter complained that federal court decisions

> which deny "Freedom of Choice," close neighborhood schools, pair classes, consolidate schools and produce forced bussing in order to reach a racial mixture in public schools are in violation of the "law of the land" and the United States Constitution. To force these unconstitutional "police state" decisions and demands upon the people is nothing less than tyranny in government by those who render such decisions.

"Who are the guilty, guilty as sin?" Maddox asked. His answer was, "Federal judges . . . who have ignored the United States Constitution." I beg of you," his letter concluded, "as a fellow American to restore 'Freedom of Choice' in education and the neighborhood schools, as guaranteed in our basic law."[29]

The Maddox letter was a voice from the southern past and, as such, hardly likely to carry weight with the Court. As the *Swann* decision was to show, the Justices had had enough of southern resistance to enforcement of the *Brown* principle. They were now ready to order whatever measures they thought necessary to end the dual school systems that still existed almost two decades after *Brown*.

6

The Justices Deliberate

Justice Jackson once said that the Court's argument begins where that of counsel ends.[1] That was certainly true when the Justices met in conference to discuss the *Swann* case. The postargument conference itself was held on Saturday, October 17. At the regular Friday conference that week, the Chief Justice had stated that the case was so important that he wanted to reserve it for a special conference on the next day.

The conference room itself was a large rectangular chamber at the rear of the Court building, behind the courtroom. One of the longer walls had two windows facing Second Street. The other, with a door in the middle, was covered with bookshelves containing reports of decisions of the Supreme Court and federal courts of appeals, as well as copies of the *United States Code* and *U.S. Code Annotated*. Along one of the shorter walls was a fireplace, above which hung a Gilbert Stuart portrait of Chief Justice John Marshall in his robes. To the left of the portrait was a forest painting by John F. Carlson. On the opposite wall hung two landscapes by Lily Cushing, a beach scene and another forest scene. The Chief Justice has said that he had these paintings hung for Justice Douglas, always noted for his love of the outdoors.

In the center of the conference room ceiling was an ornate crystal chandelier, and at one end of the room stood a table around which the Justices sat, with the Chief Justice at the head and the others ranged in order of seniority, the most senior opposite the Chief, the next at the Chief's right, the next at the Senior Associate's right, and so on. In the ceiling above the chandelier were bright fluorescent lights—another of the improvements installed by Chief Justice Burger.

The special Saturday *Swann* conference began with the usual handshakes exchanged by the Justices. The formal greetings could not, however, mask the underlying tension, both because of the public controversy surrounding the case and the fact that, as soon became apparent, the two most senior members of the Court—the Chief Justice and Justice Black—asserted a more restrictive view of the judicial power in such cases than the rest of the conference appeared willing to accept.

Burger and Black Begin the Conference

Chief Justice Burger opened the conference by saying that he considered the case so important that he wanted to dispense with the ordinary procedures. He proposed that the precedent set by his predecessor at the first conference of the Warren Court on the *Brown* case[2] be followed. Like Chief Justice Warren there, he suggested that the Justices discuss the case informally without taking any votes. Burger said that he felt it would be best just to have a roundtable discussion without any formal vote and let all air their views on the matter.

The Chief Justice then began his discussion on the merits of the *Swann* case by querying "whether any particular demands are either required or forbidden. *Brown I* said the right is a right to be free from discrimination—separation solely because of race was outlawed." This was the theme the Chief Justice was to repeat in the first drafts of his *Swann* opinion. As far as the instant case was concerned, Burger said the Court had "to look at the facts to see if there is evidence of discrimination. The rigidity of 71–29 by McMillan disturbs me. There must be some play in the joints—perhaps a 15 percent leeway?"

Justice Black, the senior Associate Justice, who spoke next, expressed the strongest conference view against the McMillan order. The Alabaman began by asserting, "It's foolish to think this question will be solved in our own or our children's lifetime. We fought a Civil War over treatment of the Negro." As Black saw it, the case "boils down to a very simple proposition. . . . what was clear was that there was to be no legal discrimination on account of race by any government."

To Black, it was plain that the courts had no business doing what Judge McMillan had done. "I have always had the idea," the Justice declared, "that people arrange themselves often to be close to schools. I never thought it was for the courts to change the habits of the people in choosing where to live." The Justice then turned to the post–Civil War amendments, saying, "The slavery amendments allow us only to enforce the prohibition against denial of equal protection." The prime responsibility here, in Black's view, lay with Congress. The Fourteenth Amendment, he pointed out, gave Congress power to enforce its prohibitions. "Policies in preventing discrimination are therefore not for us but for

Congress. If a state wants to get rid of buses and the Federal Government disagrees, let Congress provide the buses."

Douglas, Harlan, Brennan, and Marshall Disagree

The next three speakers at the conference sharply differed with the views expressed by Burger and Black. The June 27 memorandum by Justice Douglas[3] had shown plainly where he stood on the issues before the Court. His conference presentation followed his memorandum approach. "I don't think," Douglas asserted, "the issue is too involved." He focused on whether the busing ordered by Judge McMillan was proper. "Is there a prohibited discrimination in busing?" The answer depended upon the remedial power of the district court: "The problem is what is the power of the court without the help of Congress to correct a violation of the Constitution."

Douglas pointed to the broad remedial powers of the courts in other areas. "If there is an antitrust violation, we give a broad discretion." The same should be at least as true here. Nor could it be claimed that compulsory busing was itself an invalid invidious discrimination. "Is it invidious," the Justice rhetorically asked, "to take certain steps to remedy discrimination?" To Douglas, as he had put it in his memo, "bussing is a permissible tool for ending desegregation" and "the question of the precise need for bussing in a particular community is singularly appropriate for determination by the District Court." Hence, he concluded at the conference, "we ought to let the district court decide how best to disestablish" Charlotte's dual system.

Justice Harlan, who spoke next, was normally the most reserved of the Justices. But he displayed unaccustomed emotion as he sought at some length to show his agreement with Judge McMillan's order. As he saw it, "Busing is only one facet of a much more complicated problem." The real question was, "What are the basic principles" to deal with the broader problem?

Before answering this question, Harlan stressed the crucial stage that had been reached by the desegregation process. "This stage," the Justice asserted, "is almost as important as the stage the Court was in at the time of *Brown I. Brown I* settled that state enforcement of racial discrimination was a denial of equal protection. *Brown II*[4] on implementation crossed the bridge to giving the courts the job of enforcement—not Congress under section 5 [of the Fourteenth Amendment]."

Until now, Harlan pointed out, "we have largely left the matter to the lower courts, stepping in only where we became concerned with delay." With this case, however, the situation was different. "Now we must pronounce standards to guide the implementation process." Harlan then made the following points, to be used as guides to the Court's decision:

1. There is no problem of "state action" here, since "these are conventional 'state action' cases—state enforced segregation." In addition, "I would lay aside whether 'state action' is involved in housing patterns, [although] it's relevant to whether the School Board is proceeding in good faith."
2. "Given the duty to disestablish created by *Brown II,* that duty includes the duty to mix the races."
3. "There is no constitutional requirement of racial balancing and I don't think McMillan proceeded on that theory. It was simply a legitimate point of departure, like that approved by Hugo in *Montgomery.*"[5]
4. "Busing is not an impermissible tool."
5. "The neighborhood school is not a constitutional requirement if departure from it is necessary to disestablish a segregated system."

Harlan concluded by reiterating the need for the Court to lay down standards to guide the lower courts: "We won't help this problem if we're fuzzy about it." As far as he was concerned, "I start from *Green's*[6] implied principle that schools must be disestablished." This meant that the Court should "reject the court of appeals 'reasonableness' standard" and affirm Judge McMillan.

Justice Brennan then spoke in support of the McMillan order. Brennan emphasized the point he had made in his original *Green* draft opinion but had taken out from his final opinion of the Court because of the objections of some of the Justices—that a dual school system stigmatized the black children.[7] "The only way to remove the stigma of racial separation is to achieve substantial integration." The Constitution did not prescribe a particular method for achieving this integration. It was within the district court's discretion to select the appropriate method in each case. "But there can be no doubt that where busing is the only way to achieve the required amount of integration, the district judge has the power and duty to order it."

Justice Marshall also approved of Judge McMillan's order; yet, as the only black on the Court, he spoke briefly and in a restrained manner that masked the intense feelings he must have had during the discussion. At one point, Marshall referred to the "freedom of choice" plans that had been stricken down by the Warren Court. "There is no such thing," he declared, "as freedom of choice for the Negro child in the South."

The Other Justices

Justices Stewart and White had presented their views before Justice Marshall. Both had indicated support for Judge McMillan, though not as strong as that shown by Justices Douglas, Harlan, and Brennan. Stewart began by asking, "What must be done *now?*" His answer directly contradicted the view earlier enunciated by Chief Justice Burger. "Not only desegregation," Stewart asserted, "but affirmative integration is required." That was necessary to "convert to a unitary school system."

"In these de jure situations," Stewart went on, "we must say what a court (a) must do and (b) may do to dismantle historical systems." As he saw it, "the Constitution certainly permits and may require (a) benevolent gerrymandering, taking into account race (b) absolute majority to minority transfer right for Negro children." However, Stewart thought that "no quota can be required; 71–29 was wrong."

Justice White started by saying, "I agree pretty much with Potter." He thought that "the law of remedy doesn't require racial mixing as such to correct official segregation." But it was "not impossible that Charlotte is a special background where the remedy justifies no black school and no white school." From this, White went on to mild approval of the district court decree: "I won't go so far as to say that the Constitution requires a racial mix in all schools, although maybe McMillan was right"—that is, on the facts of this case.

Justice Blackmun, the last to speak at the conference, indicated at most qualified support for Judge McMillan. He started by saying, "We should not emphasize the facts in these cases as much as the principle. Do we want to lay standards? We must." Blackmun noted that he was "much taken" with the Douglas June memorandum. Blackmun then said, "I feel much as I did when I wrote *Kemp*," referring to an opinion authored by him in 1970, when he was a court of appeals judge. That opinion had refused to approve a desegregation plan under which four schools continued to be racially identifiable and completely black. In his opinion, Blackmun had stated that busing was "one possible tool in the implementation of unitary schools."[8] Now, in the *Swann* conference, Blackmun repeated this view. He did, however, add, "Busing is only a consequence of some other decision." He also noted that he was troubled by the effect of the district court order upon "the neighborhood school idea—at least at the elementary level." At an earlier point he had affirmed that he was "prejudiced toward the neighborhood school. People buy homes to be near schools."

As already indicated, the Chief Justice had begun the conference by suggesting that the session be used to permit the Justices to air their views on the matter without any formal vote. By the time everyone was "aired out" it was 5 P.M. and the session had to be adjourned. Though no vote had been taken, it was fairly clear that there was much support for Judge McMillan. Only Chief Justice Burger and Justice Black had indicated firm opposition to the far-reaching remedy ordered for the Charlotte schools. The others had stated their approval, though some doubt had been expressed about McMillan's use of the 71–29 ratio and busing at the elementary school level—yet Justice Blackmun, one of those who voiced a doubt here, did say that he thought that "racial balancing is at least a tool."

The conference discussion showed that if a vote had been taken, a majority would probably have voted to affirm the McMillan order. At least four (Douglas, Harlan, Brennan, and Marshall) would definitely have voted that way; White would in all likelihood have done so (given his statement that McMillan may have been right on the facts of this case); Stewart and Blackmun would also have probably voted to affirm McMillan (though with modifications in his decree). At any rate, no support had been expressed by the others for the rejection of McMillan's order advocated by Justice Black and, to a lesser extent, the Chief Justice.

Douglas Civil Rights Act Memorandum

Before the next *Swann* conference, three of the Justices prepared memoranda on the case. On October 31 Justice Douglas circulated a memorandum on the legislative history of the 1964 Civil Rights Act.[9] The Charlotte-Mecklenburg School Board had relied on the proviso in that law that stated that "nothing herein shall empower any official or court of the United States to issue any order seeking to achieve a racial balance in any school by requiring the transportation of pupils or students from one school to another."

The Douglas memo tried to show that the purpose of the proviso "was to insure that the act would not be read as granting any official or court the power to cure racial imbalance caused by *de facto* segregation, and to make clear that no new authority, beyond that possessed under the Fourteenth Amendment, was conferred upon the courts."

On the other hand, Douglas concluded, the proviso did not limit the authority of the federal courts to remedy situations that were constitutionally impermissible: "No attempt was made to limit the power of the courts which they possessed under the Fourteenth Amendment."

This Douglas memo played a significant part in the *Swann* decision, since much of it was incorporated by Chief Justice Burger into the second draft of his opinion of the Court. The Chief Justice may have been led to do this by the fact that Justice Harlan wrote to him on November 3 that "the argument that the Federal Civil Rights Act of 1964 prevents disestablishment bussing" was a point that "should be encompassed in any opinion" in the case.

Harlan Draft Opinion

The Harlan letter was sent to Chief Justice Burger together with a printed *Memorandum of Mr. Justice Harlan*. The latter was written in response to a request from the Chief Justice, who had told the others after the October 17 conference that he would appreciate their views on the case in written form. Harlan's letter began, "Herewith, as you requested, a copy of the exercise which I set for myself in [the case]." Harlan also wrote, "I am not circulating copies of this memorandum generally. Brother Stewart, however, has asked me for a copy and I am sending him one."

The Harlan memo was in the form of a thirty-page draft *Swann* opinion, which strongly supported Judge McMillan's order. It started by noting that the case was concerned only with

> a historically state-created and maintained system of racially segregated public school education. Accordingly, we have no occasion in these cases to treat with a public school system that is claimed to be either the indirect product of other forms of discriminatory state action or the product of patterns of society whose existence has not been influenced by some sort of invidious state conduct.

After giving a brief history of the case, the Harlan draft declared, "We reverse that portion of the Court of Appeals decision dealing with the elementary school plan. We affirm the decision of that court insofar as it relates to the senior and junior high schools." In effect, this was a complete affirmance of Judge McMillan.

According to Harlan, analysis of the case required

> a consideration of two related questions: First, does the adoption of a non-racially gerrymandered geographic zoning plan for assigning students satisfy the school board's affirmative duty with respect to this aspect of the process of disestablishing a former state-sponsored dual school system? Second, assuming a school board can be compelled to go further than adopting such a zoning plan, is the degree of actual race mixing in the school system a proper criterion for assessing compliance by the school board with its constitutional duty to disestablish such a racially dual system?

The Harlan draft then went into the controlling principles laid down by *Brown I*. Here, the Justice stated a view of *Brown I* much broader than that of Chief Justice Burger at the conference or in his first draft *Swann* opinion of the Court. As Harlan saw it, "The policies of state government which *Brown I* condemned encompassed not merely the enactment of public laws which in terms discriminated between the races, but also the implementation of those laws through the maintenance by state officials of two separate systems of public education, one for black and one for white." This meant that the racial composition of schools might clearly be taken into account by the courts. That was true because "The racially identifiable school was both a necessary end and the ultimate expression of the state educational policies condemned in *Brown I*."

Harlan urged that the courts should take into account "the results of local governmental efforts to convert from one system of decision-making to another" in framing remedial criteria. The consideration mentioned as well as "the central role played by the racially identifiable school in the former dual system and the necessity for definite remedial criteria for measuring school board compliance with the Constitution . . . require this Court's rejection of the school board's basic contention that their affirmative duty to disestablish the former dual system is met by the adoption of a student assignment plan based on nonracially gerrymandered geographic zones."

The same considerations, the Harlan draft went on, "now lead us to approve the District Court's use in this case of a remedial criterion based on results in terms of the actual racial composition of student bodies as the test for assessing the constitutional sufficiency of a proposed student assignment plan." Judge McMillan might properly decide that "the proper remedy can only be to assign students in substantially similar proportions throughout the school system. We hold, in other words, that mathematical racial balancing may be imposed on a school system where, in the district court's informed discretion, this remedy is necessary to achieve a stable desegregation plan."

The Harlan draft went on to approve the use of busing as part of the McMillan remedy. "A school board, in discharging its affirmative duty to convert to a unitary system, can at the very least be required to utilize all of the techniques ordinarily at its disposal in implementing local educational policy," including busing, which has always been "an integral part of the Charlotte-Mecklenburg system."

Nor could it validly be argued that too much busing had been ordered by the district court. According to the Harlan draft, "The question, 'How much bussing?' is therefore a part of the larger problem of the nature and strength of the presumption in favor of the elimination of all racially identifiable schools . . . which the concept of an affirmative duty to desegregate implies."

The court of appeals had, however, disagreed with Judge McMillan on busing in elementary schools. Harlan interpreted the court of appeals ruling in this respect "as holding that the amount of bussing required by the elementary school plan was unreasonable as a *matter of law.*" On that basis, the Harlan draft stated, "we think the Court of Appeals' standard fails to meet the requirement we spelled out in *Green:* namely, that the school board must demonstrate the unfeasibility of any plan which would be more effective in converting to a unitary school system."

In determining who should be bused, the Harlan draft concluded,

> the board and the district court must realize that conversion to a unitary system is not simply one among many other legitimate goals to be pursued as a matter of local educational policy. Rather, *Brown I, Green,* and *Alexander v. Holmes County,* 396 U.S. 19 (1969), construed the Equal Protection Clause as encompassing a constitutional right to be educated in a unitary, nonracial system. The correlative constitutional duty which that right requires school officials to discharge is no different from any other constitutional duty with respect to state officials; they must do their planning and maximizing of otherwise permissible state objectives in a decision-making environment which presupposes the education of all children in schools which are not racially identifiable.

The Harlan draft, though uncirculated, played an important part in the *Swann* decision process. It put Chief Justice Burger on notice that the leader of the Court's conservative wing during the Warren Court's later years would not go along with less than a clear affirmance of Judge McMillan. This meant that the Chief Justice could not expect any support from the Justice whom he might ordinarily consider his natural ally in his effort to move the Court from the posture assumed during the Warren tenure. The Harlan opposition was ultimately to be a key factor in leading Burger to modify his first draft *Swann* opinions of the Court. Without Harlan's support the Chief Justice could not hope to secure a majority for the attempt in his draft to work a virtual reversal of Judge McMillan's broad order.

Brennan Memorandum

After certiorari had been granted in *Swann,* Justice Brennan had prepared a memorandum on the case which was not circulated. Following the October 17 conference, the Justice reworked his memo and, with some changes, sent it to the Chief Justice in response to his request for something in written form.

The Brennan memo repeated the *Brown I* emphasis on the stigmatizing effect of segregation—much as the Justice had done at the conference and in his first draft of the *Green* opinion. "Since, in my view," the Brennan memo declared, "the evil of segregation was stigma, the goal and purpose of desegregation is the elimination of stigma. A unitary school system is one whose practices as to pupil assignment, faculty assignment, school site location, facilities allocation, etc., do not stigmatize any race."

Brennan then asked, "What is necessary to eliminate the stigmatizing effect of racial separation in a formerly *de jure* segregated school district?" His answer: "The only way to remove the stigma of racial separation is to achieve substantial integration." This means

> that there must be enough mixing of the races throughout the public school system that any remaining racial separation is fairly attributable not to state policy past or present, but to other factors such as *de facto* residential segregation, physical obstacles to larger school attendance zones, etc. Only when a school board has demonstrated its good faith in racial matters by bringing about substantial racial mixing will any remaining separation be fairly interpretable as not reflecting the previous state policy of treating Negroes as inferiors.

In answering the question of what constitutes "substantial racial mixing," Brennan stated, "No one set of figures will give the answer in all situations, but some rules of thumb are possible." Then, turning to the *Swann* case, he wrote, "I read Judge McMillan in the Charlotte case as suggesting that in the context of that school board's situation 71–29 is just that—a rule of thumb—a goal. And I must say that for myself I have trouble faulting his conclusion for that particular situation."

The basic principle stated by Brennan's memo was, "The Constitution requires not merely desegregation but integration," directly rejecting the more limited approach taken by the Chief Justice. "Of course," the memo asserted, "white resistance to integration is to be given no weight whatever in determining what level of integration is practical."

Brennan next asked, "How is this integration to be achieved? The Constitution does not prescribe a particular method, and a variety can be used—including neighborhood school districts gerrymandering to promote integration, bussing, school site location, etc." It was within "the discretion . . . of the courts ultimately to select in each particular case that method which achieves the constitutionally required degree of integration in the most convenient manner. But there can be no doubt that where bussing is the only way to achieve the required amount of integration, the district judge has the power and the duty to order it."

The Brennan memo contained even stronger support for Judge McMillan than that displayed in Justice Harlan's draft opinion. Burger was now on notice that two of the most influential Justices would not be satisfied with anything less than a clear affirmance of McMillan.

The Chief Justice next scheduled a second conference, which was held on Thursday, December 3, and lasted half the day. The Justices repeated the views they had stated at the first *Swann* conference. Again no vote was taken, but the sentiment plainly favored affirming Judge McMillan. Five days after the second conference, on December 8, an opinion for the Court drafted by the Chief Justice was circulated.

7

Burger Takes the Lead—
His First Draft

"Under the Constitution and Acts of Congress," wrote Justice Douglas in an April 24, 1972 letter to the Chief Justice, "there are no provisions for assignment of opinions. Historically, the Chief Justice has made the assignment if he is in the majority. Historically, the senior in the majority assigns the opinion if the Chief Justice is in the minority."

The Douglas letter was in response to Chief Justice Burger's assignment of the opinion in a 1972 case, even though he had not been with the majority at the conference. "You led the Conference battle against affirmance," Douglas declared, "and that is your privilege. But it is also the privilege of the majority, absent the Chief Justice, to make the assignment."

The letter then referred to Burger's action a year earlier in drafting the *Swann* opinion in apparent violation of the accepted assignment procedure. "The tragedy of compromising on this simple procedure," Douglas asserted, "is illustrated by last Term's *Swann*. You who were a minority of two kept the opinion for yourself which the majority could not accept." Only after extensive redrafts could the *Swann* opinion command a Court. "After much effort your minority opinion was transformed, the majority view prevailed, and the result was unanimous."

"But," the Douglas letter declared, "*Swann* illustrated the wasted time and effort and the frayed relations which result when the traditional assignment procedure is not followed."

In his 1972 letter to the Chief Justice, Douglas gave vent to his resentment over the Burger circulation of his *Swann* draft opinion. Along with other Justices who had strongly supported Judge McMillan—particularly Brennan—Douglas saw the Burger action as a clear violation of Court procedure. As Douglas would put it in his 1972 letter, "If the

Conference wants to authorize you to assign all opinions, that will be a new procedure. Though opposed to it, I will acquiesece. But unless we make a frank reversal in our policy, any group in the majority should and must make the assignment."[1]

At the *Swann* conferences, a clear majority had indicated approval of the McMillan order. At least five—Douglas, Harlan, Brennan, White, and Marshall—had signaled solid support for McMillan. The remaining majority Justices—Stewart and Blackmun—had not opposed affirming the district court. In fact, only Justice Black and Chief Justice Burger himself had stated strong negative views on McMillan's action.

If a vote had been taken along the lines just indicated, Douglas, as the senior Justice in the affirming majority, would have had the undoubted right to assign the opinion of the Court. But no vote had been taken and now the Chief Justice took matters into his own hands and frustrated the will of the majority by taking the opinion for himself, even though he had spoken in opposition to affirmance of McMillan. More than that, as a reading of the Burger draft of December 8 showed, the Chief Justice was using his position as opinion writer to try to prevent a clear affirmance of Judge McMillan.

The Chief Justice's action posed a dilemma for those, like Douglas and Brennan, who strongly supported the affirmance of McMillan. There was no doubt that a majority of the conference could overrule the Chief Justice on this or, indeed, on any other matter. For, as Justice Brennan was once heard to remark after a particularly contentious conference, "Five votes can do anything around here."[2]

Formally challenging Chief Justice Burger by putting his action to a vote was, however, quite another matter. Such a rebuff to the new Chief Justice so early in his term would destroy his effectiveness as the Court's leader and might irreparably fragment the Court itself. In addition, it was doubtful that a majority could be secured for taking the opinion away from Burger. Justices who might be willing to vote to affirm McMillan would hesitate to vote for such a rebuff of the Chief Justice.

The Justices most disturbed by Burger's drafting of the opinion concluded that the proverbial discretion made more sense than a fruitless valor. They decided not to challenge the Chief Justice and to work instead to ensure that the Burger draft did not come down as the final *Swann* opinion.

When it was circulated on December 8, 1970, the Burger draft was accompanied by a typed *Memorandum to the Conference*, which noted, "I enclose typewritten draft of proposed opinion" in the *Swann* case. Apparently the Chief Justice saw his position in this case as similar to

Chief Justice Warren's in *Brown*, for the memo contained the following paragraph:

> I am sure it is not necessary to emphasize the importance of our attempting to reach an accommodation and a common position, and I would urge that we consult or exchange views by memorandum, or both. Separate opinions, expressing divergent views or conclusions will, I hope, be deferred until we have exhausted all other efforts to reach a common view. I am sure we must all agree that the problems of remedy are at least as difficult and important as the great Constitutional principle of *Brown*.[3]

Burger Draft

The Chief Justice's draft opinion contained twenty-seven double-spaced typed pages. Its text is reproduced in Appendix A. It began by noting that, despite the number of post-*Brown I* cases, "the Court has not resolved the specific issues raised in" this case. The draft then contained in its Part I a statement of the facts and history of the case essentially similar to that in the final *Swann* opinion. Part II summarized the *Brown* decisions and stressed the "many unanticipated difficulties" encountered in their implementation. "Nothing in our national experience prepared anyone for dealing with changes and adjustment of the magnitude and complexity thereafter to be encountered." Practical implementation confronted the lower federal courts "with a multitude and variety of problems under this Court's general directive. Understandably, those courts had to improvise and experiment without detailed or specific guidelines." Now, however, this case afforded "the first opportunity to attempt some steps toward guidelines."

More specifically, according to the Burger draft, the case presented the following essential questions:

(a) whether the Constitution authorizes courts to require a particular racial balance in each school within a previously racially segregated system;

(b) whether every all-Negro and all-white school must be eliminated as part of a remedial process of desegregation;

(c) what are the limits, if any, on the rearrangement of school districts and attendance zones, as a remedial measure; and

(d) what are the limits, if any, on the use of transportation facilities to correct state-enforced racial school segregation.

These questions were similar to those stated in the final *Swann* opinion. They were dealt with near the end of Part II of the draft. Then there was a discussion in Part III of "the essential holding of *Brown I.*" This discussion was crucial, since, as the Chief Justice put it, "a proper understanding of that holding is the predicate for any guidelines that can be formulated."

Next came the passage that proved particularly troubling to the Justices who favored a clear affirmance of Judge McMillan: "It may well be that some of the problems we now face arise from viewing *Brown I* as imposing a requirement for racial balance, *i.e.*, integration, rather than a prohibition against segregation. No holding of this Court has ever required assignment of pupils to establish racial balance or quotas."

A footnote reference gave a narrow construction to *Green*[4] (it seems "to have been read over broadly by some as a mandate for integration") and *Montgomery County*,[5] which was restricted to teachers ("it does not relate to desegregation or assignment of pupils").

The draft then contained a gratuitous assertion that the Court was not concerned with anything other than separation by race in public schools forced by governmental action:

We are concerned in these cases with the elimination of the discrimination of the dual school systems, not with the myriad factors of human existence which can cause discrimination in a multitude of ways on racial, religious or ethnic grounds. . . . The elimination of racial discrimination in public schools is a large enough burden and that process will be retarded, not advanced, by efforts to use it for broader purposes not within the power of school authorities. Too much baggage can break down any vehicle.

Nor can the Court, the draft asserted, concern itself with de facto segregation: "Our objective . . . does not and cannot embrace all the residential problems, employment patterns, location of public housing, or other factors beyond the jurisdiction of school authorities that may indeed contribute to some disproportionate racial concentration in some schools."

Part IV of the Burger draft began with a restricted view of the judicial remedial power in this type of case, which the draft contrasted with what it termed "a classical equity case"—for example, where removal of an illegal dam or divestiture of an illegal corporate acquisition is ordered. "Here, however, we are not confronted with a simple classical equity case, and the simplistic, hornbook remedies are not necessarily relevant. Populations, pupils or misplaced schools cannot be moved as simply as earth by a bulldozer, or property by corporations."

Judicial remedial powers, the draft stressed, "are not co-existensive with those of school authorities." A school authority may decide "that as a matter of sound educational policy schools should be racially balanced." Such a decision would appear proper "and although we decide nothing on this, it is difficult to see what federal challenge could be successfully asserted." The same is not necessarily true of a court. "Remedial judicial authority does not put judges automatically in the shoes of school authorities whose powers are plenary."

As far as judicial power was concerned, "the ultimate remedy commanded by *Brown II*[6] restated and reinforced in numerous intervening cases up to *Alexander*,[7] was to discontinue the dual *system*." The judicial power was limited to measures aimed at "[d]iscontinuing separate schools for two racial groups." The consequence of such measures "would be a single integrated system functioning on the same basis as school systems in which no discrimination had ever been enforced." The implication was that district courts could only act to bring about the situation that would have existed had there never been state-enforced segregation. This was virtually to give the Supreme Court imprimatur to Judge Parker's view that "the Constitution . . . does not require integration. It merely forbids discrimination"[8]—a view that the *Green* decision had supposedly found inapplicable to school cases.

The Burger draft then dealt with the four questions previously stated. As far as racial balance was concerned, there were "strong intimations" by the district judge "that the 'norm' is a fixed mathematical racial balance reflecting the pupil constituency of the system." This went too far. "Neither the Constitution nor equitable principles grants to judges the power to command that each school in a system reflects, either precisely or substantially, the racial origins of the pupils within the system." Consideration of racial composition may be "one relevant step in identifying a possible violation of the rights of a class of litigants," but "it is not the function of a court, lacking as it does the plenary policy powers of a school authority, to order the individual schools to reflect the [racial] composition of the system."

In answering the second question—Must all one-race schools be eliminated?—the Chief Justice stated, "Undesirable though it may be, we find nothing in the Constitution, read in its broadest implications, that precludes the maintenance of schools, all or predominantly all of one racial composition in a city of mixed population, so long as the school assignment is not part of state-enforced school segregation."

The Burger draft's discussion of the third question, that of the remedial altering of attendance zones, began much as does the final *Swann* opinion, except for the following sentence in the draft: "Absent a history of a dual school system there would be no basis for judicially ordering assignment of students on a racial basis." The implication here was that, outside the South, federal judicial power could not deal with the segregation problem—an unnecessary raising of the issue of de facto segregation in other parts of the country.

Turning to the specific question of altering attendance zones, the Chief Justice indicated that such a remedy might be used even if it meant assigning pupils outside their neighborhoods. But he did so in a grudging fashion, saying only, "The pairing and grouping of non-contiguous school zones is not an impermissible tool but every judicial step in shaping such zones that goes beyond combinations of contiguous areas should be closely examined." In addition, the draft asserted that any guidelines the Court could fashion would be "inescapably negative."

On the last question—transportation of students—the Burger draft stated that "affirmative action in the form of student transportation may properly be used" as a remedial device, but its use had to be limited. First of all, according to the draft, "the objective should be to achieve as nearly as possible that distribution of students and those patterns of assignments that would have normally existed had the school authorities not previously practiced discrimination." The implication once again was that the remedial goal was only the elimination of state requirements for segregation, not integration.

In determining the limitations on the use of busing, "the courts must also weigh the burdens upon students and the potential frustration of legitimate educational goals. It hardly needs expression that the limits on time of travel will vary with many factors, but with none more than the age of the students." The draft went on to acknowledge the difficulty of the judicial task in determining how much busing to order. "It calls for the wisdom of Solomon and the patience of Job; and by and large district judges have exhibited these qualities in the painful period of transition." The kudo for district judges scarcely obscured the fact that the Chief Justice had approved the narrowing of Judge McMillan's order by the court of appeals to exclude busing to elementary schools, for that action had plainly been based upon "the age of the students."

In concluding, the Burger draft referred to the court of appeals' use of the concept of "reasonableness" as a test "to define the limits on the equitable remedial power of the District Court," as well as the *Green* use of "the term 'feasible' and at least by implication, 'workable' and 'realistic.'" The implication was that the court of appeals and *Green* tests were similar.

But the Chief Justice's draft did not take the logical next step of affirming the court of appeals. Instead, its concluding sentence was, "The cases are remanded for reconsideration by the District Court and further action not inconsistent with this opinion."

The Opposition Begins to Form

The Burger draft confirmed the misgivings of those Justices who supported Judge McMillan's order. They had feared that the Chief Justice might use his position to thwart the clear affirmance that the majority favored. Now their apprehensions appeared justified. The Burger draft had gone out of the way to denigrate McMillan's integration effort, particularly in its assertion that the Constitution required only elimination of state-enforced segregation, not the fostering of integration. At the same time, it was overly conciliatory toward southern school boards, going so far at one point as to refer to their "most valiant efforts" to meet the desegregation requirements.

Even when the draft endorsed what McMillan had done, it did so in a begrudging manner, as when it said that the McMillan technique of pairing schools was "not an impermissible tool." What was worse was that, to McMillan's supporters, the entire draft was negative and indecisive in tone. The problems involved were "undecided" by prior opinions of the Court; they were due to the "unanticipated difficulties" caused by the *Brown* decisions. Consideration of pupil population ratios was "one relevant step," but racial balance could not be compelled. Only a begrudging endorsement was given to busing: it might be used, but must be strictly limited. Although the draft could be read as upholding the portions of McMillan's order dealing with junior and senior high school students, there was no specific message of approval for such judicial action.

Even more troubling to the legal technicians on the Court—especially Justice Harlan—was Burger's approach to the remedial powers of the district courts. The draft itself recognized the far-reaching authority traditionally possessed by courts of equity, conceding, "Once a right and a violation have been shown, the scope of equitable remedies to redress past wrongs is broad for in the nature of equitable remedies breadth and flexibility are essential." But then the Chief Justice drew a gratuitous distinction between remedial power in a case involving "a classic equity claim" and this desegregation case. Though equity judges have normally had power to do whatever they deemed necessary to redress a violation, here only a more limited remedial power should be approved. All the district court could do in such a case was to act to bring about a "system functioning on the same basis as school systems in which no discrimination had ever been enforced." The implication was that affirmative measures to secure integration were not approved, even though the district judge deemed them to be the only effective measures to correct the constitutional violations proved by plaintiffs.

It was scarcely surprising that the strongest supporters of Judge Mc-Millan—Justices Douglas, Brennan, and Marshall—quickly sent the Chief Justice objections to his draft. What was perhaps unexpected by Burger was the negative reaction of Justice Stewart, who had been much milder in his approval of McMillan at the conference.

Douglas Letter

Justice Douglas was the first to respond to the Burger draft. In a December 10 letter to the Chief Justice, he listed his objections: *"First.* You seem to intimate that racial balance is required by the District Court order." Douglas wrote that he disagreed with such an interpretation. "I read the District Court's opinions to mean that an all-white or all-black school is not *per se* improper." Douglas did not interpret McMillan's order as *requiring* racial balance but, to the contrary, as saying that one-race schools might be permissible, depending on the facts of each case.

"*Second,*" Douglas wrote, "You seemingly make irrelevant 'imbalances resulting from residential patterns of the area,' even when those patterns were created by state law." Douglas again stated that he disagreed. "It is difficult to remove from *de jure* segregation any type of state action that was a cause of separate school systems." Douglas also rejected the Chief Justice's limitation of the *Montgomery County* decision[9] to teachers. According to Douglas, "The distinction between teachers and students would seem relevant on the issue of feasibility, but not on the issue of constitutional power to order integration."

The main Douglas objection was to the Burger draft's limited conception of busing. "As I read your opinion, transportation of students is limited to attaining the 'distribution of students and those patterns of assignments that would have normally existed had the school authorities not previously practiced discrimination.'" As Douglas saw it, "If this is the only 'discrimination' that can be cured, the orders for integration would seem to be quite limited." This meant that "a school could be built in the ghetto, or in the white neighborhood, if numbers required, and need not be integrated."

The district court would have only the task of determining if any schools were built for the purpose of racial segregation. "If so," Douglas concluded, "the District Court would determine where schools would have been built, absent such a motive, construct attendance lines around these areas, and bus students to the schools in those areas which they would have attended. This does not seem likely to result in Blacks attending suburban schools or Whites attending schools in the central city. It would only integrate the central city areas."

Marshall Draft

Justice Marshall was the next to be heard from. On December 22, he wrote to the Chief Justice, "I have carefully gone over your draft and finding myself unable to deal with it section by section I have drafted a statement of my views in the form of an opinion." A copy of the draft was enclosed; it was circulated to the other Justices on January 12, 1971.

The Marshall *Swann* opinion was, as the Justice acknowledged in his December 22 letter, "a very rough draft." Consisting of almost seventeen typed pages, it contained a clear affirmance of Judge McMillan's order. Over half of the draft (some ten pages) was devoted to a statement of the facts and the history of the case. Marshall then declared flatly, "the time has come for the era of dual school systems to be ended." To accomplish that result, a broad view of the judicial remedial power was needed: "when school boards fail to meet their obligations it is up to the courts to find remedies that effectively secure the rights of the Negro children." The "traditional . . . principles of equitable discretion" were invoked. "And implicitly the equally traditional doctrine of appellate review that recognizes the necessity of flexibility in equitable decrees and that required a strong showing of abuse of discretion to reverse such a decree was also invoked."

"Here the District Court's conclusion," Marshall wrote, " 'that all the black and predominately black schools are illegally segregated' was clearly supported by the record." In such a situation, the district court had full discretion to take any measures it deemed appropriate to end the dual system: "in a case such as this where so little was done under this Court's command of 'all deliberate speed,' to fail to make every effort to formulate a feasible plan that would eliminate such schools would be to deny Negro students in Charlotte the relief they had been waiting on so long."

In Marshall's view, "it was clearly within the District Court's equitable power to set out broad goals to guide efforts to work out such a decree." The use of the 71–29 ratio as a guideline "was clearly not an abuse of discretion." The same was true of the district court's consideration of "the demographic patterns of the community," as well as "the influence of non-school board governmental action on these patterns. It would be a hollow remedy indeed for a[n] equitable court to provide a remedy that removed all traces of direct school board action designed to maintain the dual school system and to ignore other governmental action, which dramatically affected the schools."

In addition, it was not an abuse of discretion for Judge McMillan to conclude that it was necessary to modify the grade structure to require children to go to schools not near their homes. "Very early in the development of public schools courts recognized that a student had no right to attend the school nearest his home that served his grade." In this case, "The District Court's conclusion that assignment of children to the school nearest their home serving their grade would not produce an effective dismantling of the dual system was supported by the record."

Turning to the busing aspect of McMillan's order, the Marshall draft found that "the techniques used in the District Court's order were certainly within that Court's power to provide equitable relief." The busing ordered was "well within the capacity of the school board." Most of the required buses "were already available and the others could easily be obtained." The transportation ordered was no more onerous than that already provided in Charlotte.

The Marshall draft declared that the McMillan plan "clearly reflects . . . the sound tradition of equity" and met the *Green* requirement of a remedy that promised realistically to work and to work *now*. "While such a solution may not be necessary to remove the effects of school discrimination in other places, we cannot hold that it was an abuse of discretion for the District Court to find it appropriate in Charlotte-Mecklenburg." Accordingly, the Marshall draft concluded, "although we would have reversed the portion of the Court of Appeal's [*sic*] judgment requiring a remand, the order of the District Court after remand reinstating the original decree is now before us and we affirm that order. Affirmed."

Brennan's Objections

The strongest criticism of the Burger draft opinion was contained in a December 30 letter sent to the Chief Justice by Justice Brennan. The Justice, who had spoken forcefully in favor of affirming Judge McMillan at the *Swann* conferences, began by declaring, "The approach taken in your draft . . . differs considerably from that which I believe is required."

In the first place, wrote Brennan, "I cannot accept the implications of your statement" that the Court had "not resolved the specific issues" raised in the case. On the contrary, the Justice asserted, "I feel that the cases following *Brown*,[10] particularly *Green* v. *County School Board*,[11] travelled far down the road, so that all we are really required to do here is fill in the outline constructed by *Green*."

Brennan wrote that he was particularly concerned with the negative thrust of the Chief Justice's draft, which "should I think be positive." The Justice objected to the Burger statement "that 'perhaps any guidelines are inescapably negative.' That assertion conflicts with our approach in *Green*—indeed with our approach in all previous cases—and, I respectfully submit, is wrong."

The Brennan letter urged that the Court set *positive* guidelines:

> I believe that it is not only possible, but imperative, that we continue, as in *Green,* to formulate positive guidelines, and had thought we took these cases to carry-on (in perhaps a more detailed and specific way) what we began in *Green.* In my judgment, therefore, we should set out guidelines as specifically and positively as possible and give further content to them by applying them in a detailed consideration of the facts in the present case. This would provide necessary guidance to the District Courts in dealing with these very difficult and complex problems.

The Justice then listed the points that should be made in the *Swann* opinion:

1. "*Green* . . . held (1) that the system of segregation must be dismantled and its effects undone; (2) that the goal of this process was the abolition of racially identifiable schools; and (3) that all feasible steps must be taken to achieve this goal."
2. The district court considered all the elements of a unitary system listed in *Green.* Under its decree, "effective steps have now been taken to produce a unitary school system." The Court should affirm the district court's action "as action necessary to disestablish a dual system."
3. "The constitutional command as developed in our cases is to eliminate *now* the stigmatizing effect of racial segregation in a formerly *de jure* segregated school system. The only way to dismantle a dual school system and remove the stigma of racial separation is to achieve substantial integration."
4. "In achieving the disestablishment of a dual school system, the remedial power of a federal court is broad. Compulsory transfers for that purpose violate no constitutional right of either black or white children."

Here Justice Brennan took direct issue with the Burger draft's restricted conception of judicial remedial power. As Brennan saw it, "when reviewing the orders of district courts in this area, we are confronted wholly with a problem of the propriety of equitable remedies to redress the constitutional violation of operating an officially segregated system."

In this case, according to Brennan, "we ought to make explicit our approval of the following techniques in disestablishing a dual school system":

"(A) It is constitutionally permissible for either the court or the school board to take race into account in assigning pupils in order to dismantle a segregated system," and this "ought to be stated squarely in this case."

"(B) One acceptable remedial measure is the establishment as a goal or rule of thumb—not an inflexible quota—the attainment in each school of a racial mix equivalent to the racial mix of the entire system." *Montgomery County* applies to students as well as teachers. "In the present case, I would simply follow the *Montgomery* formulation and hold that racial balancing of student bodies is a permissible remedial tool to facilitate desegregation."

"(C) To attain the mixing of the races required to disestablish a dual system, a variety of techniques including gerrymandered zones, satellite zones, pairing and clustering of schools, compulsory transfer plans, and majority-minority transfer plans, are permissible." All of the techniques used by Judge McMillan were constitutionally permissible.

"(D) Bussing is only an incident of the remedial techniques itemized above and should not be viewed as a separate issue which can be analyzed on its own terms." It may be "almost impossible to provide firm guides" for busing. "Nonetheless, I believe we should take a major step in that direction by holding that, in the circumstances of this case, Judge McMillan's order was appropriate since it involved an expenditure of less than one percent of the school district's operating budget and since the average bus trip was well within the statewide average."

"(E) A majority-to-minority transfer plan should always be part of a desegregation plan since it provides important play in the system's joints and may act as a deterrent to re-segregation."

The Brennan letter concluded by urging deletion of the Burger draft's reference to other than official segregation. "I feel that an attempt to decide the *de facto* situation is unwarranted by the facts of this case and suggest that the discussion [in] your draft be reserved for a time when we are squarely faced with this problem."

Stewart Memorandum

The conference discussions had prepared the Chief Justice for the opposition displayed by Justices Douglas, Brennan, and Marshall. It is less likely that he expected Justice Stewart to join the opponents of his draft *Swann* opinion. Stewart had been considered a moderate since his appointment to the Court in 1958. He had never acted on the basis of a defined judicial philosophy such as that which guided the activists like Douglas and Brennan.

During the Warren years, Justice Stewart had been something of a "swing man," voting at times with the Chief Justice and his activist allies and at others with Justices Frankfurter and Harlan, who led the advocates of judicial self-restraint. In the main, however, Stewart tended to vote more with the Frankfurter-Harlan view than with the Court's activist wing. As Judge Learned Hand put it in a letter to Frankfurter about Stewart, "He really does not believe himself vested with authority to direct the fate of 200,000,000 people through the 1st, 5th, and 14th Amendments."[12] Unlike the Justices on either end of the Court's polar extremes, Stewart approached cases on their individual facts, without trying to fit them into the gestalt of a consistent philosophy. This tended to give him the reputation of a moderate, though one more likely to follow the doctrine of judicial restraint than the increasingly activist approach of the Warren-led majority.

At the *Swann* conferences Stewart had been more restrained than some of the others in expressing approval of the district court order. This and the Justice's normally middle-of-the-road stance led the Chief Justice to count on Stewart's backing. The Justice had, however, become convinced after the conferences that what Judge McMillan had tried to do was deserving of more generous support. Stewart told me that one of the first things he did after the first *Swann* conference was to look McMillan up in *Who's Who*. The Justice had never heard of the district judge and he wanted to find out what kind of person he was. He found that he and McMillan were about the same age and had similar backgrounds. McMillan had attended law school at Harvard, while Stewart had gone to Yale and both had been in the navy in World War II. Stewart felt immediate empathy with the district judge and looked for ways to back his action.

The Justice decided to try his hand at a draft opinion in order to refine his own thinking on the case. Stewart was a judge who believed that if an opinion in support of a decision did not write well, the reasoning behind it must be fallacious. He wanted to see if his own leaning in favor of McMillan could be supported by a written analysis. He decided to label his draft as a memorandum, rather than an offered opinion disposing of the case. Any other approach would have constituted a direct rebuff to the Chief Justice, since the Justice had not been assigned the opinion in the case.

Stewart had finished his own draft opinion when the Burger first draft was circulated. The Justice was disturbed when he read the Chief Justice's draft opinion. He had come to the conclusion that the Court should deal with Judge McMillan's order as generously as possible, and the Burger draft appeared to him to be ungenerous.

On December 14, a few days after he received the Burger draft, Stewart sent the Chief Justice a letter which he termed "a preliminary response to your circulation of December 8." Stewart wrote, "With a great deal of what is said in your proposed opinion I can fully agree. I do, however, have serious reservations about your treatment of some of the problems."

The principal disagreement noted in Stewart's letter was that

> you purport not to decide the constitutionality of "a school authority decision that as a matter of sound educational policy schools should be racially balanced on the declared premise that this is desirable or necessary in order to prepare children for the obligations of citizenship in a pluralistic society." I think it is important to state that such a school board decision would be wholly constitutional.

In addition, Stewart wrote that he considered "it a *sine qua non* of a decree that it provide for a compulsory majority to minority transfer at the option of the transferor." Stewart also said that he agreed that *Swann* should "be remanded for reconsideration."

More important than his letter to the Chief Justice was the copy of his own draft that Stewart included with the letter. "In an effort to straighten out and organize my own thinking . . . ," Stewart wrote in his letter, "a good deal of time over the last several weeks has been devoted to the preparation of the enclosed memorandum. I have not circulated it generally, and shall not do so until after you have had an opportunity to consider it fully."

The Stewart memo was a complete draft opinion containing 38 printed pages. Headed "Memorandum of Mr. Justice Stewart," it began by posing the questions presented: "When a school board has adopted a plan designed to remedy a violation of *Brown* v. *Board of Education*,[13] and that plan is challenged in a United States district court, what standards should the district judge apply to test the plan's adequacy? What objectives may he set, and what means may he employ, in ordering changes in such a plan?"

Stewart noted that under *Alexander* v. *Holmes County Board of Education*,[14] school boards were obliged to terminate dual systems at once and to operate only unitary schools. "Partly as a result of that decision, we are now confronted with the problem of defining in more precise terms than have heretofore been necessary the kind of action by a local board that meets the requirement of immediate conversion to a 'unitary' system."

The Stewart draft then asserted that it was "clear . . . that no individual child has a substantive constitutional right to attend a school having any particular racial mixture. The fact, without more, that a child living in a school district containing children of different races attends a school in which his race is in the majority—or the minority—does not of itself establish a violation of any constitutional guarantee." There also had to "be a showing of official action of the State designed to create or maintain a situation of racial isolation." In this case, the state-action issue presents no problem, since "there has been a deliberate, knowing attempt on the part of school officials vested with state power to maintain a dual system."

The crucial question presented is that " of permissible and appropriate remedies for the clear violation of the defined constitutional right." In dealing with it, "we must keep our role clearly in mind." *Brown II*[15] allotted to "the district courts . . . the primary responsibility for 'elucidating, assessing, and solving' the problems presented."

The key factor here, Stewart emphasized, is that the judicial power "was an *equitable* power that was to be in the *discretion* of the district judge." The Stewart memo took direct issue with the Burger notion that judicial remedial power in such a case was somehow narrower than in traditional equity cases. "A school desegregation case," Stewart asserted,

> does not differ in these respects from any other case involving the framing of an equitable remedy to repair the denial of a constitutional right. The task is to correct, fully, effectively, and with a fair regard to all the individual and social interests represented, the state of affairs which offends the Constitution, and to make the correction in such a way as to reduce to a minimum the chances of a recurrence of the wrong.

The 1964 Civil Rights Act, in Stewart's view, did not limit judicial authority in this respect.

The Stewart memo then turned to the school board's argument "that the basic meaning of *Brown v. Board of Education* was a requirement of 'colorblindness' in the administration of public school systems." This meant "that action, whether by a local board on its own initiative or by a district court within its remedial discretion, which takes race into account in remedying a violation of *Brown I* is self-contradictory because [it is] mandating exactly what the decision prohibits."

Referring to this board argument, the Stewart memo categorically asserted, "This reasoning is fallacious." In the first place, said Stewart (repeating the point made in his letter to the Chief Justice), school boards have broad power and "It is within their power to decide that the establishment in each school of a ratio of Negro to white students equal to that for the district as a whole best meets the educational or social objectives of the community." But that does not mean that a board discharges its affirmative duty to desegregate simply by following "the rule of colorblind neighborhood zoning [under which] each child attends the school having space for him which is located nearest to his home." Stewart flatly "reject[ed] the proposition that a school board found to have maintained a dual system discharges its affirmative duty by establishing a system which does not 'take race into account.'"

As Stewart saw it, "Viewed as a remedy for decades of self-imposed segregation, colorblind neighborhood zoning (supposing the phrase to have a precise meaning) is closely analogous to the 'freedom of choice' plans that were before us in . . . *Green.*" In *Green*[16] the Court had found that the freedom-of-choice plan had not made for any change in the dual system. Here, too, color-blind neighborhood zoning had proved inadequate. "I would hold that where, as here, it appears that 'color blind' neighborhood zoning would result in a pattern of school attendance essentially similar to that which existed under the dual system to be disestablished, it is not enough to meet the affirmative remedial duty of the local board."

The Stewart memo next turned to the mechanics of desegregation. It dealt briefly with other aspects of the process and then focused on what it termed the "central issue" in the case, that of student assignment. Here Stewart stressed again that he "rejected the contention of the local boards that as a matter of substantive constitutional law they are *prohibited* from doing more than establishing an assignment system based on 'color blind' neighborhood zoning," as well as "the argument that the equitable remedial power of a United States district court is limited to the imposition of such a zoning scheme." On the contrary, "A district judge may, in framing a remedy under *Brown II,* take into account the consequences of his action for the racial composition of the schools in the system."

The district court should, of course, take conflicting interests into account in framing a remedy. And "as among *remedies,* it is of course incumbent on the judge to choose that which vindicates the right at the least possible cost to other public and private interests."

As an "obvious first step," a majority-to-minority transfer plan should be required in all desegregation plans and that was provided by the lower courts. In addition, "it was also clearly correct for the district courts to order that the school boards make student-assignment decisions in such a way as to achieve a higher degree of actual desegregation than would have resulted from 'color blind' neighborhood zoning." The Stewart draft specifically approved the principal means employed by the district court: "benevolent racial gerrymandering" of zone lines, noncontiguous zoning, and school pairing or grouping by adjustment of grade structure and zone boundaries. "Each is therefore an appropriate part of . . . the district judge's inventory of means to the end of disestablishing the dual system."

Stewart then dealt with the busing issue: "I come now to the argument that whatever the status of a particular technique, its use is impermissible if it results in the transportation by bus of children who would otherwise have walked to school." The widespread use of busing, apart from any desegregation plans, was noted, both in the country and in Charlotte-Mecklenburg itself. In this case, most of the needed buses were already available. "In these cricumstances, I can find no basis whatever for a holding that a local school board may not be ordered to employ bussing as a method of school desegregation. Desegregation plans cannot be limited to the walk-in school." Busing, like the other techniques ordered by Judge McMillan, "should be available to local boards and district courts in pursuit of the goal of disestablishment."

The Stewart memo summed up what had been said by the principle "that the goal of local boards and district courts in fashioning remedies under *Brown II* should be to use the techniques available to achieve the greatest possible degree of desegregation in fact, taking into account the practicalities of each of the means employed." Stewart specifically rejected the test that a desegregation plan was adequte only if it eliminates every "racially identifiable school. . . . In rejecting this proposed test, I would also reject *a fortiori* the notion that a desegregation decree must require that every school in the system must have a student racial composition commensurate with that of the student population of the entire district."

The Stewart memo also rejected the notion that the district courts should seek to eliminate the effects of housing discrimination: "it is to place far too heavy a load on a school desegregation decree to ask that it also be a housing desegregation decree. It is not incumbent on the district judge in a school case to alter or even to remedy the effects of patterns of housing segregation." Under this approach "it should be clear that the existence of one or more all-Negro, or virtually all-Negro schools within a district is not in and of itself the mark of a system which still practices segregation by law." The courts should not follow any *"per se* rule of results."

The Justice concluded his memo by stating, "I would affirm those parts of the Court of Appeals' opinion which upheld the District Court's plans with respect to junior and senior high schools." As for the elementary schools, the memo noted that the means used by Judge McMillan in redrawing the school board's plan "so as to achieve a greater degree of actual desegregation were altogether appropriate means. None of them was in itself offensive to the Constitution, and so far as appears on this record, none of them was used in such a way as to contravene the principles I think should guide the equitable discretion of a district court."

Nor were "the results in terms of actual desegregation achieved by Judge McMillan through application of the various techniques . . . in themselves beyond the power of a district judge acting in the exercise of his equitable remedial discretion." In ordering the elimination of "every all-Negro school" and reassignments of Negro children to white schools, McMillan had acted within his powers.

Having gotten to the point where a full affirmance of Judge McMillan appeared forthcoming, the Stewart memo drew back in its last paragraphs. The memo ended by stating that both the court of appeals and district court judgments "should be vacated, and the case remanded for further consideration." That should be done, Stewart wrote, because "I am unable to determine, after careful examination of the record, whether or not Judge McMillan, in applying legitimate techniques to achieve a result which is not in itself constitutionally impermissible, considered himself to be bound by a standard different from that which I have set out above."

Stewart explained that his uncertainty arose from the passages in Judge McMillan's order which said that he had tried "to reach a 71–29 ratio in the various schools" and ordered the elimination of all-black schools and pupil assignment so that all grades "have about the same proportion of black and white students." On the basis of such passages, Stewart wrote, "I have come to the conclusion that Judge McMillan may have felt compelled in modifying the Board's desegregation plan to adopt a 'no racially identifiable school' test" similar to that which Stewart had already rejected. According to Stewart,

> such a test is not the criterion of compliance with our *Brown II* decision, since such a test prevents full consideration of the conflicting interests which must be reconciled in an equitable remedial decree. For this reason, and not because I find any defect in either the means used or the particular results achieved, I would vacate the judgment and remand the case to the District Court.

Stewart's lame ending can be explained by the Justice's desire to avoid a direct rebuff to the Burger draft. Stewart tried, so far as possible, to keep away from a frontal conflict with the Chief Justice. It was, after all, not his job to lead the opposition in the case. Despite this, the opinion in Stewart's memorandum was wholly unlike the *Swann* draft opinion that the Chief Justice had circulated. Except for its conclusion, the Stewart memo had come down squarely on the side of vigorous action to end dual school systems. The Justice urged an expansive conception of judicial remedial power, directly contrary to the Burger restricted notion. He stressed the discretion vested in the district courts and rejected the view that they could only impose "a requirement of 'colorblindness' in the administration of public school systems"—a view to which the Burger draft had all but given its imprimatur. Most important, the Stewart memo approved both the means used by Judge McMillan (including busing) and the results achieved by him. Even with its weak ending, the Stewart memo amounted to all but an affirmance of McMillan and, as such, a direct challenge to the Chief Justice's draft opinion.

8

Burger Second Draft

The Chief Justice was now on notice that his *Swann* draft opinion was opposed by a majority of the Court. Justices Douglas, Marshall, Brennan, and Stewart had written to him, directly disagreeing with the Burger draft, and Justice Harlan's memorandum, which had been sent to the Chief Justice well before the latter had circulated his draft, had taken a position plainly incompatible with the Burger restrictive posture. Of the other three Justices, Burger could count only on the firm backing of Justice Black. Justices White and Blackmun had both expressed support (though less strongly than some of the others) for Judge McMillan at the conference and both had indicated approval of a broader remedial power in the federal courts than the Burger draft was willing to concede.

Most significant was the opposition to the Chief Justice's negative approach by Justices Harlan and Stewart. If the Chief Justice thought he could count on anyone besides Black and Blackmun (who, as seen in Chapter 2, voted so closely with Burger when he first came on the Court that the two were popularly known as the Minnesota Twins), it was Harlan and Stewart. If he could hold neither the leader of the Court's conservative wing since Justice Frankfurter's retirement nor its leading moderate, he knew that he could scarcely expect to secure a majority for his draft. Only by rewriting the draft to meet the objections that had been expressed could he hope to obtain a majority for his opinion.

The Revised Draft

Despite this, the Chief Justice's efforts to meet the criticisms directed at his draft opinion were almost as grudging as his approval of Judge Mc-

Millan had been. On January 11, 1971, Burger sent around a revised draft of his *Swann* opinion. It was accompanied by a *Memorandum to the Conference,* which stated, "Except for the first 8 pages the enclosed draft is revised throughout. I have tried to take into account the views expressed by several of you in memos sent in response to my request. Obviously, some points were in conflict with my own position. Subject to this, I remain open to further comments and suggestions."[1]

The second draft ran thirty-seven typed pages. Its first eight pages, which were unchanged from the first draft, still contained the statement that the "Court has not resolved the specific issues raised in" this case as well as the summary of the facts and history of the case that had begun the first draft. Despite the suggestion to the contrary in the Chief Justice's covering memorandum, Part II of the draft, which summarized the *Brown* decisions,[2] was also essentially unchanged from the first draft. It still referred to the "many unanticipated difficulties" that were encountered in the implementation of the *Brown I* principle and continued to assert that "nothing in our national experience prepared anyone for dealing with changes and adjustments of the magnitude and complexity to be encountered." The only difference in Part II was the following affirmative statement in the second draft: "Cases in the district courts and courts of appeals in the past few years make plain that we should try to develop some guidelines for school authorities and courts." In addition, the four questions that Burger said were presented in the case, which were at the end of Part II in the first draft, were now moved to a later part of the opinion.

Part III of the Burger revised draft began with the assertion that "it may well be that some of the problems we now face arise from viewing *Brown I* as imposing a requirement for racial balance, *i.e.*, integration, rather than a prohibition against segregation. Indeed the term integration nowhere appears in any opinion dealing with pupil segregation and no holding of this Court has ever required assignment of pupils to establish a fixed racial balance or quotas." This was virtually the same as a statement in the first draft that had been so troubling to the Justices who favored a strong affirmance of Judge McMillan. And, at this point, the second draft still contained the footnote which gave a narrow construction to *Green*[3] and restricted *Montgomery County*[4] to desegregation of faculty. It also contained the gratuitous assertion that the Court was not concerned with anything other than the elimination of school segregation—not with "the myriad [other] factors" which can cause other types of discrimination. The statement that the Court could not concern itself with de facto segregation was also repeated from the first draft.

Part III of the second draft concluded with a new paragraph:

It is axiomatic, but perhaps deserves restatement since it is so often over-looked, that the judicial function in cases of this kind is not broadly to shape the most socially desirable ends or determine policy but to enforce the Constitution. And it is not only a Constitution we are expounding but a *written* Constitution that puts limits on courts as well as others. In policy and program the authority of the political branch—Congress, the states and school authorities—is broader than that of courts; the former are the responsive and responsible policy makers. Our task is to deter-mine what is commanded and what is prohibited by the Constitution. Much that a majority or even all of this Court might consider desirable and proper lies beyond our power to command and we serve that Con-stitution best if that is our guide.

But this passage—for all its pastiche of the famous John Marshall state-ment[5]—merely reinforced the Burger view that the remedial power of the courts in this area was somehow less than that in others, and notably less than that of the school authorities. This only reemphasized the failure of Part IV of the second draft to broaden the restrictive view of judicial power which had been taken in the Chief Justice's first draft.

It is true that Part IV of the revised draft toned down somewhat the restrictive language in the first draft on the courts' remedial authority. The contrast of a school case such as *Swann* with "a simple classical equity case," with its corollary that judicial power was somehow broader in the latter case, was eliminated. Instead, the draft now read, "A school desegregation case does not differ fundamentally from any other case involving the framing of an equitable remedy to repair the denial of a constitutional right. The task is to correct, with a fair regard to all the individual and collective interests the condition that offends the Con-stitution."

Yet the draft still indicated that, somehow, the judicial power was not as clear-cut as in "a classical equity claim," like one involving the "wrong-ful erection of a dam." There, "the remedy is simple: equity will com-mand the offending dam to be removed." Regarding the case at hand, the Chief Justice still contended that the situation was different. Repeat-ing the language of his first draft, he wrote, "Here, however, we are not confronted with a simple classical equity case, and simplistic hornbook remedies are not necessarily relevant. Populations, pupils or badly located schools cannot be moved as simply as earth by a bulldozer."

Again Burger stressed that judicial remedial powers "are not co-exten-sive with those of school authorities" and that the judges are not placed in the shoes of school authorities whose powers were plenary. The clear implication was that the federal courts did not possess powers as broad as those of school boards in remedying violations of the *Brown* principle.

The draft did, however, expressly follow the suggestion made by Justice Stewart in his December 14 letter, that the Court indicate that a school board decision that schools should be racially balanced "would be wholly constitutional."[6] The Burger second draft specifically stated:

The relevant commands of the Constitution are negative prohibitions, but it is wholly within the power of school authorities to conclude that, given the particular circumstances and needs of the community, to prepare students to live in a pluralistic society, that the establishment in each school of a ratio of Negro to white students and faculties equal to that for a district as a whole best meets the overall educational objectives of the community.

But the Chief Justice went on to stress that school boards alone had such powers. As to the judiciary, he wrote, "A federal equity court, however, has no such roving, at-large powers."

Part IV of the Burger second draft now concluded with two pages on the argument "that the equity powers of federal district courts have been limited by Title III of the Civil Rights Act of 1964." The draft used much of the language of Justice Douglas's October 31 memorandum in rejecting this argument.

The Burger revised draft contained a new Part V (taken almost verbatim from the Stewart memorandum), which discussed aspects of desegregation other than problems of student assignment. Here, the draft summarized the *Green* and *Montgomery County* cases and stated that school boards had the "remedial responsibility . . . to eliminate racial discrimination" with respect to matters such as transportation, extracurricular activities, faculty and other school personnel, and construction and closing of schools. "When necessary, district courts should retain jurisdiction to assure that these responsibilities are carried out."

In its Part VI the revised Burger draft began with the statement contained in Justice Stewart's memorandum, "The central issue is that of student assignment." With regard to it, the Burger draft went on, "four basic questions are presented in this case." Then the draft repeated the four questions stated in Part II of the first Burger draft:

(1) whether the Constitution authorizes federal courts to require a particular racial balance in each school within a previously racially segregated system;

(2) whether every all-Negro and all-white school must be eliminated as part of a remedial process of desegregation;

(3) what are the limits, if any, on the rearrangement of school districts and attendance zones, as a remedial measure; and

(4) what are the limits, if any, on the use of transportation facilities to correct state-enforced racial school segregation.

In his December 30 memorandum,[7] Justice Brennan had criticized these questions because "the thrust of the phrasing . . . is also negative and should I think be positive." The Justice suggested a rephrasing of the questions in positive terms:

> For example, I'd phrase the questions: (a) whether the Constitution permits courts to require a mixture of whites and blacks in each school within a previously racially segregated system; (b) whether any all-Negro and all-white schools may be retained as part of a remedial process of desegregation; (c) whether the rearrangement of school districts and attendance zones, as a remedial measure, must be effected; (d) whether the use of transportation facilities to correct state enforced racial school segregation is constitutionally permissible.

Now, in his second draft, the Chief Justice rebuffed this Brennan suggestion and retained the phrasing he had used in his first draft.

In answering the first question, that of racial balance, the revised Burger draft virtually repeated the language of the first draft. Again the Chief Justice treated Judge McMillan's use of the 71–29 ratio as requiring racial balance (quoting in greater detail from the district judge's order on this point) and indicated that such a requirement might not be imposed, declaring again, "It is not the function of a court, lacking as it does the plenary policy powers of a school authority, to order the individual schools to reflect the composition of the system."

The second question, that of one-race schools, was also answered by the second Burger draft much as it had been in the Chief Justice's first draft. Again, the conclusion was that, "undesirable though it may be," there is nothing in the Constitution to preclude the operation of schools "all or predominantly all of one racial population," so long as it was "not part of state-enforced school segregation." There was also an express approval of optional transfer plans (a point Justice Stewart had insisted on in his December 14 letter): "an optional majority-to-minority transfer plan is an obvious and necessary part of every desegregation plan."

With regard to the third question, remedial altering of attendance zones, the revised draft was also substantially similar to the Chief Justice's first draft. Again the statement was made, "Absent a history of a dual school system there would be no basis for judicially ordering assignment of students on a racial basis." Again this was an unnecessary raising of the de facto segregation issue, with the implication that the federal courts could not deal with the segregation problem outside the South—an implication that troubled even the normally conservative Justice Harlan (who wrote an "X" next to the statement on his copy of the revised Burger draft).

It is true that the revised draft did eliminate the double negative which had lamely approved the pairing and grouping of noncontiguous school zones: "is not an impermissible tool" in the first draft was changed to "is a permissible tool." But this minor concession was more than outweighed by the inclusion, just before the change just noted, of the passage (contained in the section on busing in the first draft) limiting the remedial power of the federal courts: "The objective should be to achieve as nearly as possible that distribution of students and those patterns of assignments that would have normally existed had the school authorities not previously practiced discrimination." Again the Chief Justice had stated his negative view that the remedial goal was only the elimination of state requirements for segregation, not integration. (Well might Justice Harlan write, "X?" on his copy of the revised draft next to this passage.)

In addition, the Chief Justice repeated his statement that "any guidelines are inescapably negative," adding this time "as indeed the Equal Protection Clause is negative." Again Justice Brennan had been rebuffed, this time in the objection in his December 30 letter to only negative guidelines.

The discussion of the last question posed, that of student transportation, started out much as the first draft did, by asserting that the scope of permissible transportation could not be defined with precision and that "no rigid guidelines can be given." But the second draft then stated a more positive attitude toward busing, which was declared "a normal and accepted tool of educational policy." Ordering of busing was expressly found to be within the remedial power of the district courts: "we can find no basis for a holding that local school authorities may not be ordered to employ bus transportation as a method of school desegregation. Desegregation plans cannot be limited to the walk-in school."

The force of this clear affirmation was, however, lessened by a stress in the next paragraph of the revised draft upon limitations on the busing power. Once again the Chief Justice declared that there might be legitimate objections to busing, repeating from his first draft, "It hardly needs expression that the limits on time of travel will vary with many factors, but with none more than the age of the students." As in the first draft, the clear implication was that the court of appeals had been correct in its narrowing of Judge McMillan's order.

The first draft's two concluding paragraphs were now contained in a separate Part VII of the revised draft. Once again the Chief Justice referred to the court of appeals' "reasonableness" test and indicated that it was similar to that laid down in *Green*. Once again, too, the draft ended with a remand to Judge McMillan.

Before the remand order, there was a new paragraph on the resegregation problem: "At some point in the near future, these school authorities and others like them should have achieved full compliance with this Court's decision in *Brown I.*" Then, anticipating the logical response such a statement would engender, namely, that the communities involved may not necessarily remain demographically stable, Burger ended the paragraph by noting that desegregation decrees are not appropriate for judicial monitoring of demographic changes or shifting residential patterns. "Neither school authorities nor district courts are constitutionally required to make year by year adjustments of the racial composition of student bodies once the affirmative duty to desegregate has been accomplished and racial discrimination through official action is eliminated from the system." Only if there is "a showing that either the school authorities or some other agency of the State has deliberately attempted to fix or alter demographic patterns in order to effect racial separation in the schools, would . . . further intervention by a district court" be permissible.

Then came the same concluding paragraph as in the first draft: "The case is remanded for reconsideration by the District Court and further action not inconsistent with this opinion."

Maneuvering for Affirmance

To the majority that favored affirming Judge McMillan, the Chief Justice's revised *Swann* opinion was hardly an improvement over his first draft. In his January 11, 1971, *Memorandum to the Conference* transmitting the second draft, the Chief Justice had stated that, except for its beginning, the draft had been "revised throughout" and that, in making the revisions, he had "tried to take into account the views expressed by several of you." The revised draft itself belied these assertions. Whatever changes had been made were minor and, though two of the suggestions made by Justice Stewart had been included, they also were relatively unimportant. Virtually nothing had been conceded to those who had written in support of Judge McMillan's order.

Most important, the revised draft still retained the negative tone that both deeply offended McMillan's supporters (notably Justice Brennan) and suggested an attempt on the Court's part to conciliate the school authorities (once again their efforts were termed "most valiant"). The view taken of judicial remedial power remained restricted and the approval given to the remedies ordered in the case grudging and hedged in by limitations. Instead of the affirmance of Judge McMillan that the majority desired, the case was still remanded to him for reconsideration of his decree.

Now began the maneuvering for affirmance by those Justices who most strongly approved of Judge McMillan's action. But first there were two formal responses to Chief Justice Burger's second draft. The day after it was sent around, Justice Marshall circulated the opinion which he had sent to the Chief Justice on December 22, 1971. The Marshall draft, as we have seen, affirmed McMillan and was, as such, a formal rejection of the new Burger draft.

Douglas Draft Dissent

The Chief Justice's revised draft also spurred Justice Douglas into circulating a dissenting opinion, which was largely based upon the June 27, 1970, memorandum that the Justice had written. Circulated on January 13, the fourteen-page printed opinion was headed, "Mr. Justice Douglas, dissenting in part." The Douglas dissent began by rejecting the argument that school assignments based on race were unconstitutional: "Is it possible that school assignments based on race are 'invidious' when their purpose is not to downgrade an individual but to undo a wrong previously done by a State when it created a segregated school system? I think not. If there is discrimination in that setting it is 'benevolent' or 'remedial,' not 'invidious.'"

Douglas asserted that black schools were inferior, because over "the school boards are white boards, not boards representing a cross-section of the community." Blacks could not escape their inferior schools without busing. Where the "school board may be niggardly and refuse to provide an adequate bussing service. . . . [t]he federal court should have power to enter a decree requiring bussing. Otherwise the State gives the black no real choice but to stay in his *de jure* segregated school. That is denying access to the quality school by reason of race."

The courts must also have power "to put blacks in white schools and whites in black schools—not in numbers that inundate a school and resegregate it, but in numbers that achieve a disestablishment of the segregated school." That alone will "make equal protection a reality."

The Douglas draft dissent then stated, "My differences with the Court, except for emphasis, are in the main threefold." First, the Justice asserted, "federal courts are authorized, within limits, to require so-called racial balance to rectify the conditions resulting from *de jure* segregation of public school systems." But, noted Douglas, even the contrary assertion did not require reversal of Judge McMillan. "For as I read the several opinions below, I conclude that no such effort was made in the present case. The District Court merely took the 79% to 21% [*sic*] ratio as a guideline, and no more." Such a "rule of thumb" was permissible under

Montgomery County, "Where we approved faculty assignments precisely on that basis."

Second, wrote Douglas, he saw no occasion to discuss de facto segregation. "We have not heard argument on this issue and should not decide that important question without full-dress consideration." And third, "I think we travelled most of the distance we need to go today in *Green v. County School Board.*"

Under *Green,* Douglas affirmed, the district judge might use all the tools he did to disestablish Charlotte's dual school system. The school board's "neighborhood school" theory meant only a perpetuation of segregation. "There is much talk both legalistic and sentimental about the values of the neighborhood school." People may normally like their neighborhood school." People may normally like their neighborhood. "But the victims do not. They do not like subhuman housing; they do not like higher prices or usurious interest rates; they are not at all attracted to their overcrowded, understaffed, decrepit schools." Judge McMillan was right in his assertion that the quality of a child's education should not depend on the racial accident of his neighborhood.

In particular, Douglas commented, the district court had the power to find "that the bussing provisions of its order as regards elementary schools were necessary to eliminate racially segregated schools in the Charlotte-Mecklenburg school district." District courts have wide remedial discretion in these cases. "The task of a district court," the Douglas dissent concluded, "is to disestablish a *de jure* segregated school system. And the question of the precise need for bussing in a particular community is singularly appropriate for determination by the District Court. I cannot say that that court was so far out of bounds that mark an abuse of discretion."

Pressures on Stewart

The Douglas draft dissent could have served as the *Swann* opinion of the Court if it had been joined by the majority who favored affirming Judge McMillan's order. But Justice Douglas himself realized that his opinion was too one-sided to command the broad support needed. In addition, he recognized that he was scarcely the person to serve as the leader of a unified opposition to the Chief Justice in so controversial a case. Douglas was known among his colleagues as the maverick on the Court. He was an idiosyncratic loner, who was least effective in the give-and-take required in a collegial institution.

In many ways, the Douglas antithesis in this respect was Justice Brennan, who had served as the catalyst for some of the most important decisions of the Warren Court. After Justice Frankfurter's retirement, Brennan had become the most active lobbyist among the Justices, always willing to take the lead in trying to reconcile differences in opinion so as to mold a majority for the decisions that he favored.

Brennan agreed with Douglas that the latter was not the man to lead the opposition to the Chief Justice's *Swann* drafts. At the same time, he expressed the view that he too was not the best person to organize a movement to secure an affirmance of Judge McMillan. It would be better if the leadership could be assumed by a Justice not associated so closely with the Warren Court's activist stance.

At this point, both Douglas and Brennan obtained copies of the long draft opinion which Justice Stewart had sent to the Chief Justice on December 14. After they had read his draft, the two Justices decided to have a private conference with Stewart in his chambers. The meeting was held on Tuesday, February 16, and, according to a Justice, "was the first real break in the case."

The two Justices told Stewart that they agreed with the standards he set out in his draft. Then they tried to persuade him that Judge McMillan had in fact followed those very standards. In his draft opinion, Stewart had relied on McMillan's use of the 71–29 ratio to reach his conclusion that the case should be remanded to the district judge. Brennan and Douglas stressed that, in his draft, Stewart had read McMillan as setting forth the 71–29 ratio as not merely a goal, but rather as an inflexible requirement. They persuaded him that that interpretation was incorrect and that the correct approach was that taken in the Douglas draft dissent: "The 79–21 [*sic*] ratio was used as a guide, not an absolute."

Even more important in its effect upon Stewart was the perspective in which Brennan and Douglas placed the case. They persuaded Stewart that if McMillan were affirmed, this case, both because of its facts and the scope of the district court order, could have a very significant impact upon the de jure segregation that still existed. Stewart finally agreed to prepare a redraft affirming McMillan and emphasizing the portions of the district judge's opinions stressed by Brennan and Douglas, which tended to show that the percentages were used as guides only.

But Justices Brennan and Douglas still believed that something more substantial was needed. Someone was needed to seize the initiative from the Chief Justice by writing an opinion giving effect to the majority's view of the case and presenting it for a vote. Douglas telephoned Brennan to discuss the matter. He reaffirmed that he felt that he was not in a position to take the lead, but that Brennan might well succeed in the matter. Brennan replied that he thought that Stewart was the one to lead the way, since he had a fully written opinion that, with the changes he had agreed to make, would obtain the votes of Brennan, Douglas, and probably Marshall, Harlan, and maybe White as well.

Stewart was, however, as a law clerk put it, "more than reluctant." It was clear that the Justice would not take the lead in confronting the Chief Justice. Stewart did, however, prepare the redraft of his opinion that he had promised.

Stewart Draft Dissent

The Stewart redraft was prepared and printed soon after the February 16 meeting among Justices Douglas, Brennan, and Stewart. Consisting of 33 printed pages, it was headed, "Mr. Justice Stewart, dissenting in part," and dated "February _____, 1971." The crucial difference between this draft dissent and the draft opinion contained in the memorandum the Justice had sent to the Chief Justice earlier was in the ending, which, instead of remanding (as the draft in the memo had done), concluded with a clear affirmance of Judge McMillan.

Apart from this, the two drafts were virtually similar. There were, it is true, minor differences. Thus, in stating the principle that no child had a constitutional right to attend a school with a particular racial mixture, the Stewart draft dissent added, "But every school board does have a substantive constitutional duty to operate a wholly desegregated school *system.*"

In addition, in dealing with the problems of "resegregation," the draft dissent repeated the Stewart memo's statement that it was appropriate for the district court to consider those problems. It then added:

> The danger of resegregation may justify the requirement of a higher degree of actual desegregation of particular schools than would be necessary given a likelihood that populations will remain stable. Where a decree creates a disparity between schools or between areas which invites the migration of white parents from one school zone to another, it does not adequately perform its function of disestablishing the dual system.

More important was a statement added regarding the impact of different desegregation techniques on schoolchildren. Now Stewart noted:

> As far as concerns junior and senior high schools, these consequences of desegregation techniques are likely to be relatively insignificant. Schools at the secondary level tend to be large and draw students from considerable distances. The age of the students reduces the alarm of parents that school reorganization will be harmful. Bussing or the use of public transportation [is] inevitable for most students in any case, and [has] long since been accepted.

The Justice did, however, concede that at the elementary school level "the situation is less clear." The Stewart dissent then touched on the argument that "at least at the elementary level," the attendant "costs" of desegregation were such as to render the techniques used by Judge McMillan constitutionally impermissible. Justice Stewart went out of his way here to declare, "I would reject this *per se* position as analytically untenable."

Then, summarizing the different desegregation techniques used by Judge McMillan, the Stewart dissent asserted, "Each, with its advantages and disadvantages, must be available to local boards and district courts in pursuit of the goal of disestablishment." (The Stewart memo had used the weaker "should be" instead of "must be.")

As already noted, the chief difference between the two Stewart drafts was in their endings. Both drafts began their concluding sections by affirming "those parts of the Court of Appeals' opinion which upheld the District Court's plans with respect to junior and senior high schools." The draft dissent, like the memo, also rejected "a rule of 'racial balance,' by which every school in a district must have a racial composition commensurate with that of the district as a whole, or a test, by which no school may deviate by more than a set percentage from the racial proportion of the district." But, unlike the memo, the draft dissent noted that such per se rules were invalid. "I would feel obliged to vote to remand this case for further proceedings if it were clear to me that Judge McMillan had adopted one or the other of them. I have concluded, however, that on a fair reading of the record before us no such doctrinaire understanding can be imputed to him."

In accordance with the Brennan-Douglas suggestion, Stewart now stressed the portions of Judge McMillan's opinions which indicated that racial balance and the 71–29 ratio were not an inflexible requirement. After quoting the district judge on this, Stewart wrote, "Over the 18 months of litigation . . . Judge McMillan repeatedly and insistently expressed his view that 'racial balance' was neither compelled by the Constitution nor in any other sense a requirement of a legally adequate desegregation plan."

The Stewart draft dissent quoted again from Judge McMillan—this time a statement that "racial balance" was not required under the Constitution, nor were all-black schools in all cities unlawful, nor must all school boards bus children. In light of this, Stewart declared:

> It is clear to me that Judge McMillan adopted as his goal the maximum possible amount of actual desegregation given the practicalities of the situation. I think it would be a mistake to conclude that he adopted any constitutionally derived *per se* rule of 'racial balance,' or that he disre-

garded any of the conflicting interests which it was his duty to consider and reconcile. In this context, he took the 71/29 ratio of white to Negro students in the schools as no more than a starting point for the process of planning, rather than as an inflexible requirement. From that starting point he proceeded to a final decree which was fully within his discretionary power to frame, as an equitable remedy in the particular circumstances that confronted him.

Then came Stewart's concluding sentence: "I would therefore affirm the District Court order of August 3, 1970, and remand these cases to the Court of Appeals for further proceedings consistent with this opinion."

The stage was now set for a confrontation between the Chief Justice and those who favored a clear affirmance of Judge McMillan. In the latest Stewart draft the latter had an opinion that could easily be converted from a draft dissent into the opinion of the Court. At least five, and probably seven, Justices were in agreement with the views so carefully presented in Justice Stewart's draft. If he persisted in the restricted approach taken in his two drafts, Chief Justice Burger now risked losing his Court and having his effort to control the *Swann* opinion completely frustrated.

9

The Chief Justice's Turnabout

On February 19, 1971, Justice Stewart sent copies of his draft dissent to Justices Douglas and Brennan. Accompanying them were handwritten notes. The one to Douglas read, "Dear Bill—The attached was rather hurriedly prepared after our conversation with W.J.B. on Thursday, and before I learned of later developments. Although it now may never be circulated, I thought you should have a copy for perusal at your leisure—in confidence." Stewart added, "P.S. I've also sent a copy to Bill Brennan."[1]

The reason for the statement that the Stewart dissent "now may never be circulated" was occasioned by the fact that, as a law clerk puts it, "the C.J. evidently got wind of the developments, mainly that there was rapidly developing a solid Court for affirming McMillan."

To understand what happened here, one must distinguish between the Supreme Court as it appears to the outside world and to those who participate in its decision process. The Court has always had a deserved reputation for preserving the confidentiality of its nonpublic proceedings. As far as the press and the public are concerned, the postargument stage in the Court is completely closed. The sole knowledge those outside the Court have about a case after oral argument is obtained from the announcement of the decision and the opinions filed by the Justices. Only those privy to the Court's internal workings are aware that the published opinions may not give anything like a true picture of the Court's decision process.

Internally, however, the Court building is a hotbed of gossip and rumor. Among the law clerks particularly, there is constant scuttlebutt that keeps the clerks and their Justices abreast of the latest developments in the give-and-take between the Justices that is a crucial element in their decision process. The Court's rumor mills may not grind slowly, but they do grind faithfully. Through them, Chief Justice Burger was soon made aware of the redrafting of the Stewart memorandum and the threat it posed to his own leadership in the *Swann* case.

When the Chief Justice learned that a majority was starting to form behind the Stewart redraft, he realized that he would have to change his own draft substantially or lose hope of being the *Swann* opinion writer. The first thing to do, of course, was to head off the Stewart redraft and prevent it from becoming the opinion of the Court. To accomplish this, Chief Justice Burger himself went to the Stewart chambers. He told the Justice that he had independently concluded that Judge McMillan had to be affirmed. He also said that he would use more of Stewart's language in his next draft of the opinion of the Court.

Justice Stewart's reaction was described by the same law clerk: "P.S. now felt that he had been completely boxed in and that if the C.J. circulated a draft affirming he would have to join." When Stewart told Justices Brennan and Douglas what had happened, they were equally disturbed. All three vowed that they would never again be led into submitting separate drafts to the Chief Justice without copies to the conference.

Harlan Letter

Chief Justice Burger's turnabout on affirming Judge McMillan was also influenced by Justice Harlan, with whom Burger had privately discussed the case. On February 16, Harlan sent a letter to the Chief Justice which firmly supported McMillan's order. The Justice wrote that, under the Court's decisions, racial "mixing is a permissible, if indeed not a required *remedial* tool for the disestablishment of state-enforced dual school systems." Harlan stated that the Burger revised draft seemed to blur this basic point.

The Justice's disagreement with the Chief was, however, more fundamental:

> My basic trouble with your opinion . . . is that it does not come to grips with either the district court's opinion or that of the court of appeals. I continue to think that the district court handled the matter correctly and that its judgment should be affirmed, of course with explication of our reasons. I also still think that the court of appeals was wrong in its "reasonableness" test, and that the criticism of its reasoning is a matter of substance and not mere terminology as your draft suggests. I therefore think that its judgment, insofar as it remanded the case to the district court for reconsideration under the "reasonableness" formula, ought to be reversed.

The Harlan letter made it clear to the Chief Justice that the Justice would not go along with anything less than a categorical affirmance of Judge McMillan. As already stressed, without the support of both Justices Harlan and Stewart, it would be impossible for Burger to muster a majority. The only way to win the two Justices to his side and keep control of the *Swann* opinion was for the Chief Justice to yield on the crucial issue of affirming McMillan.

Burger Third Draft

Chief Justice Burger, we have seen, had told Justice Stewart that he had decided to prepare a new draft affirming Judge McMillan. The Chief Justice circulated the third draft of his *Swann* opinion on March 4. The draft was not accompanied by any covering memorandum and consisted of 35 typed pages.

Part I of the third draft (eight pages) was taken verbatim from the second draft. It repeated the prior draft's statement that the Court had not resolved the specific issues raised in the case. Part II, containing a summary of the *Brown* decisions,[2] was also essentially unchanged. Again the statement was repeated that "nothing in our national experience prepared anyone for dealing with changes and adjustments of the magnitude and complexity encountered," as well as the statement that the Court should try to develop guidelines, though this time with the qualification "however incomplete and imperfect they may be."

Much of the introductory language of Part III was omitted from the third draft, particularly the statement that *Brown* does not impose a requirement of "integration, rather than a prohibition against segregation," and that the Court has never "required assignment of pupils to establish a fixed racial balance." But the force of the omission was lessened by the express assertion that "The objective today remains what it was in 1954—to root out all state-imposed segregation in public schools. That was the evil struck down by *Brown I,* as violative of the Equal Protection guarantees of the Constitution. That was the violation sought to be corrected by the remedial measures of *Brown II.*"

The third draft also eliminated the Part III footnote that had given a narrow construction to *Green*[3] and limited *Montgomery County*[4] to faculty. The footnote had supported the omitted statement quoted above on pupil assignment to secure racial balance. This opened the way to the statement in Part IV, to be noted, that Judge McMillan and the court of appeals had only followed *Montgomery County* in deciding this case.

Also left out of the third draft was the passage on the inability of the courts to enforce desirable social policies. Instead, the new Part III went directly into the remedial power of the federal courts (Part IV of the second draft). Part III now began by repeating the statements on the breadth of equitable remedial power and on the school desegregation case not differing from other cases involving an equitable remedy. Deleted was the comparison with "a classical equity action" that had been contained in the first two drafts. Once again, however, the statements and omissions were more than offset by repetition of the assertion that the remedial powers of the courts were somehow less than those of school boards. There were slight changes in language, but the third draft was essentially the same as the earlier drafts on this point, specifically repeating that "it is beyond judicial authority [to] establish in each school a fixed ratio of Negro to white students equal to that for the district as a whole."

Part III of the third draft concluded with the two pages that had been added to the second draft rejecting the argument "that the equity powers of federal district courts have been limited by Title III of the Civil Rights Act of 1964." The new draft's Part IV (dealing with aspects of school desegregation other than student assignment) was taken almost verbatim from the prior draft. The one difference of note was the omission of the second-draft statement "There is no reason to reconsider our holding in *Montgomery*." Instead, there was only the statement "The principles of *Montgomery* have been followed by the District Court and the Court of Appeals in this case."

Part V of the third Burger draft began (as did Part VI of the second draft), "The central issue is that of student assignment." It then repeated the "four basic questions" exactly as they had been stated in the earlier draft.

The discussion of the first question, racial balances, began with the second draft's statement, repeating the claim that the district judge had imposed a fixed 71–29 ratio and intimating that his " 'norm' is a fixed mathematical racial balance reflecting the pupil constituency of the system." Then came some new material, which began, "The Constitution, of course, does not command integration; it forbids segregation. Thus a federal district court is without power to require, as a constitutional matter, that any particular racial balance or ratio be permanently maintained." This statement was, however, immediately qualified: "that does not mean that as a *corrective* for past constitutional violations an awareness of the racial composition of the whole school system may not be an appropriate starting point in shaping a remedy."

Here the district court found that a dual system had been maintained and that the school board had defaulted in its duty to come forward with an acceptable plan: "these findings are abundantly supported by the record. It was because of this total failure of the school board that the District Court was obliged to turn to Dr. Finger to do for the Court what the board should have done."

The new draft concluded its discussion on this point by again noting, "Some of the language of the District Court can indeed be read as intimating that racial balance is a constitutional requirement." But this did not require invalidation of Judge McMillan's order: "reading that language in context and relating it to the result we conclude the District Court was addressing itself to remedies and for that purpose we cannot say it abused its discretion in formulating the remedy." In effect, this was to adopt the approach Justices Douglas and Brennan had urged, and which Justice Stewart had adopted in his draft dissent.

The new draft's discussion of the second question, one-race schools, repeated that contained in the previous draft. The same was true of the discussion of the remedial altering of attendance zones. These portions of the third draft still referred to the "most valiant efforts" of the school boards and contained the statements indicating that the remedial goal in these cases was only to eliminate state discriminatory requirements, not to achieve a unitary school system, and that the federal courts could not deal with the segregation problem "[a]bsent a history of a dual school system."

On the other hand, the new draft did omit the statement that any guidelines laid down would have to be "inescapably negative." The Chief Justice had finally eliminated the passage to which Justice Brennan had strenuously objected.

The third draft's discussion of the last question, student transportation, was also taken almost verbatim from the Chief Justice's prior draft—with two important additions. In the first place, there was now an express statement upholding Judge McMillan's busing order: "The District Court's conclusion that assignment of children to the school nearest their home serving their grade would not produce an effective dismantling of the dual system is supported by the record. Thus the remedial techniques used in the District Court's order were within that court's power to provide equitable relief."

In addition, though the section on busing still concluded with the statement that reconciliation of the competing values of granting a remedy for violation of constitutional rights, on the one hand, and the well-being of children, on the other, was a difficult task, the new draft added, "but fundamentally no more so than remedial measures courts of equity have traditionally employed."

Part VI of the new draft was also essentially the same as the concluding section of the previous draft. But, here too, there were two crucial changes. After setting forth the court of appeals' "reasonableness" and the *Green* "feasible" and "workable" tests, the draft expressly stated, "On the facts of this case we are unable to conclude that the order of the District Court is not reasonable, feasible and workable."

Then, after repeating the prior draft's paragraph on the resegregation problem, the Chief Justice's third draft contained a new concluding paragraph (instead of the remand to the district court at the end of the two earlier drafts): "The judgement of the Court of Appeals is affirmed as to those parts in which it affirmed the judgement of the District Court. The order of the District Court dated August 7, 1970 is also affirmed."

The Opposition Wavers

The Chief Justice's third draft gave those who had been most dissatisfied with his earlier *Swann* drafts what they wanted most, an express affirmance of Judge McMillan. In other respects, however, the new Burger draft was not that much of an improvement over the previous versions. To the Justices who were strong supporters of McMillan, the draft was still unduly negative. It described the problem presented to the district court as unresolved and unanticipated and still referred to the "most valiant" efforts of the school boards. It went out of its way to assert, "The Constitution, of course, does not command integration; it forbids segregation." This restatement of Judge Parker's supposedly repudiated view[5] was now even more express than it had been in the earlier drafts. This was made still clearer by the repetition from the earlier drafts of the statement "The objective should be to achieve as nearly as possible that distribution of students and those patterns of assignments that would have normally existed had the school authorities not previously practiced discrimination."

What the Chief Justice had done was to repeat the essentials of his two previous drafts and then change the ending to affirm Judge McMillan. To be sure, he had eliminated some of the language that had disturbed the others, but even his action in that respect had been qualified by his retention of other passages from the earlier drafts and his inclusion of the new Parker-like assertion referred to in the preceding paragraph. It is true that he had added language indicating that the district court's rejection of the neighborhood school concept was supported by the record and that the McMillan order met the tests laid down by the court of appeals. (Yet, even here, the force of the last statement was weakened by the double negative used: "On the facts of this case we are unable to conclude that the order of the District Court is not reasonable, feasible and workable.")

All in all, the Chief Justice had not changed the substance of his draft that much, but had merely done an about-face on the draft's conclusion. Yet this was really what the opponents of the earlier drafts wanted most—an express affirmance of Judge McMillan's entire order. The way the opinion was written certainly remained important to them, but the crucial thing was to affirm McMillan clearly. That would send a plain signal to the lower courts that would not be obscured by even the fuzzy opinion that was evolving from the Burger drafts.

Douglas and Stewart Respond

One matter that troubled some of the Justices about the third Burger draft was that it still indicated negative answers to the Court's ability to deal with de facto segregation and even some forms of de jure segregation not directly attributable to the school board. Justice Douglas, who was the first to respond to the third *Swann* draft, focused upon this point in his March 6 letter to the Chief Justice.

Douglas objected to a paragraph Burger had added in his third draft to his discussion of racial balance which stressed that the Court's objective was only school segregation: "it does not and cannot embrace all the problems of racial prejudice in residential patterns, employment practices, location of public housing, or other factors beyond the jurisdiction of school authorities, even when those patterns contribute to disproportionate racial concentrations in some schools."

Douglas wrote that this "paragraph excludes from *de jure* segregation relevant to school problems both restrictive social convenants and racial public housing." Douglas thought "that such state-sanctioned practices are included in *de jure* segregation for purposes of the public school problem." The Justice asked the Chief Justice "not [to] decide the scope of *de jure* segregation . . . and restrict *Swann* to the case where there had been a dual school system." If that were done, Douglas concluded, "Then I could join *Swann*."

The Douglas indication that he was ready to join the Burger opinion if the reference to restrictive covenants and public housing was eliminated came as a surprise to those like Justice Brennan who were still trying to bring about more substantial changes in the *Swann* opinion. Douglas had, however, come to the conclusion that a clear affirmance of Judge McMillan was more important than the opinion written in the case. The language used would be less significant than the message to the lower courts in Supreme Court endorsement of McMillan's order.

It will be recalled that after the Chief Justice had told Justice Stewart that he had concluded that Judge McMillan should be affirmed, Stewart had indicated to Justices Douglas and Brennan that he would have to join a Burger affirming draft. Now that the Chief Justice had circulated such a draft Stewart did agree to join. On March 8, he wrote a "Dear Chief" letter: "I join your opinion for the Court in this case, subject, of course, to consideration of such modifications as others may suggest."

Stewart told Brennan that his joining was only conditional. He said that the reservation in his note left him complete freedom to leave the Chief Justice if he did not sufficiently heed the suggestions of the other Justices.

The Stewart note to the Chief Justice also contained an important suggestion: "Since time is beginning to run short, I trust we can have a conference to discuss the final disposition of this case . . . very soon, hopefully this week."

The intimation by Stewart that a conference was needed struck a responsive chord among the Justices. Those who had been most offended by the tone of the Burger drafts thought that a conference was long overdue, as indeed it was. No vote had ever been taken in the case. Indeed, since the conference on December 3, the Justices had not met to discuss the case. Never, since the Chief Justice's first draft, had the Justices conferred together on the matter.

The Stewart suggestion did lead to the scheduling of a conference on Swann, which was held on March 18. Now the Justices who opposed the Burger drafts could confront the Chief Justice and secure an opinion in accordance with the will of the conference majority. But the Douglas and Stewart letters showed that the opposition to the Burger drafts was beginning to waver. Now that the Chief Justice had changed his conclusion to one of affirmance of Judge McMillan, both Douglas and Stewart had indicated that they might be tiring of the battle and be willing to join the Burger opinion. Others too might decide that the symbol of affirmance was more important than the deficiencies they still perceived in the Chief Justice's draft.

Brennan Objections

One who was not willing to give up the effort to improve the Burger draft was Justice Brennan. On March 8, he sent a six-page letter to the Chief Justice outlining his continuing objections to the latest Burger draft. Brennan's letter began, "At my request Potter gave me a copy of his revised memorandum." Brennan noted that both Stewart and the Burger third draft reached the same result. "But the approaches of the two opinions differ significantly, and I hope we may discuss the importance of the differences at the forthcoming conference."

"For me," Brennan stressed, "the matter of approach has assumed major significance in light of signs that opposition to *Brown* may at long last be crumbling in the South. . . . Although affirmance of Judge McMillan's judgment should of itself further the process, I nevertheless suggest that our opinion should avoid saying anything that might be seized upon as an excuse to arrest the trend."

The Justice then indicated his clear preference for the Stewart opinion: "Some things said in your third circulation seem to me to present that hazard. Potter's draft on the other hand seems to me not to present it."

The Brennan letter repeated the criticism made in the Justice's December 30, 1970, letter of the Burger statement that the Court had not resolved the specific issues raised in the case, pointing out again "that all we are really required to do here is fill in the outline constructed by *Green*." Brennan asserted "Your thrust that we are confronted with 'unresolved issues' appears again in several places." On the other hand, "in contrast, Potter's draft enumerates the questions for decision in the very first paragraph and . . . characterizes them as simply questions calling for application of principles spelled out in the earlier cases."

The difference here was of great significance, Brennan continued. "Potter thus gets across the message so important to emphasize in this delicate area—that this case announces no new principles but simply requires decision how school boards which have maintained segregated school systems are required to apply the principle settled as far back as *Brown II* to convert 'to a unitary system.' "

Brennan also objected to the Burger draft because it "strikes me as expressing a sympathy for these local boards that I don't think is warranted. We deal here with boards that were antagonistic to *Brown* from the outset and have been noteworthy for their ingenuity in finding ways to circumvent *Brown's* command, not to comply with it."

The Chief Justice's conciliatory approach to the school boards was unjustified. "I think any tone of sympathy with local boards having to grapple with problems of their own making can only encourage continued intrans[i]gence. I find none of that in Potter's draft. On the contrary, local boards are firmly told that they must get on with the job and I think this is not only right but that it is critically necessary to say so."

The Brennan letter also criticized those of the Chief Justice's statements indicating that the judicial remedial power was less extensive than that of the school authorities. "Those statements may be read, it seems to me, as suggesting that District Courts have less authority in fashioning remedies than we held in *Green* that those courts have." The Stewart draft did not contain any comparable limiting statements. "I think that this is preferable because any intimation that the *Green* principle has been restricted is bound to produce unfortunate consequences."

Brennan then strongly opposed the Burger statement that the Constitution does not command integration, but only forbids segregation:

> I think this would be a most unfortunate statement for us to make at this juncture in the struggle to gain compliance with *Brown*. That statement in almost *haec verba* ['the same words'] was the rallying cry of the massive resistance movement in Virginia, and of die-hard segregationists for years after *Brown*. It calls to mind Judge Parker's opinion which caused so much trouble for so long a time.[6] To revive it again would I think only rekindle vain hopes.

The Brennan letter concluded with an animadversion against "the negative approach" of the Burger draft: "I repeat that I think we might court a revival of opposition if we provide slogans around which diehards might rally." Brennan recognized "that my concern is primarily with the tone of your circulation. But as our experience with 'all deliberate speed' proved, tone is of primary importance."

At the end of his letter, Justice Brennan warned that he was prepared to vote against the Chief Justice's opinion: "I may say that I am quite prepared to join Potter or an opinion which follows his lead." This was a clear warning that the unanimity of the Court behind a Burger opinion might be broken if the Chief Justice did not further modify his draft.

The force of the Brennan warning that he might join Justice Stewart's opinion had, however, been weakened by Stewart's own note joining the Chief Justice's opinion, as well as by Justice Douglas's indication that he was close to doing so. Without the support of Douglas and Stewart, the Brennan criticisms could scarcely be expected to derail the Burger opinion, particularly if the Chief Justice made further changes demonstrating his willingness to meet some of the objections still raised against his draft.

Harlan Letter

At this point, Justice Harlan directly intervened in the *Swann* drafting process. On March 11 he sent a seven-page "Dear Chief" letter, which contained "my specific suggestions for further revision. Subject to what may eventuate from these suggestions, I am prepared to join your opinion."

In the first place, Harlan objected to the passage in the Burger draft in which "you define the remedial objective for the district courts and school boards as one of achieving 'as nearly as possible that distribution of students and those patterns of assignments that would have normally existed had the school authorities not previously practiced discrimination.' " As Harlan saw it, "that standard asks the school boards and district courts to imagine the pattern of community life that would have emerged if historically there had never been dual systems and then devise desegregation plans accordingly. I feel that this standard cannot offer any real guidance."

"To remedy this problem," Harlan proposed the omission of the paragraph containing the offending statement. In his copy of his letter, Harlan wrote next to this suggestion, "Not omitted." However, while it is true that the Chief Justice did not omit the paragraph from his next draft, it was substantially modified in the fifth (the first printed) *Swann* draft. In particular, that draft eliminated the sentence on achieving only the student distribution and assignment patterns that would have existed had there been no enforced segregation.

Harlan also proposed a new paragraph to be added between the two paragraphs that made up Part V(2) of the Burger draft, dealing with one-race schools, where the Chief Justice had said nothing in the Constitution precluded such schools. The proposed paragraph stated that "the need for remedial criteria of sufficient specificity" warranted a presumption against one-race schools and required close scrutiny of such schools by reviewing courts in order to assure that they were unrelated to resistance to desegregation. As will be seen in the next chapter, the Chief Justice did add the new paragraph, but, as contained in his fourth draft, the paragraph combined language taken from both Justice Stewart and the Harlan letter.

Harlan made a number of other suggestions that were followed by Burger. He urged the Chief to take out the sentence "The detail and nature of these dilatory aspects are not now relevant," asserting, "I think . . . the delaying tactics of this school board are relevant to the District Court's exercise of its remedial discretion." He also suggested removal of the phrase "no one could foresee in 1954" in referring to the post-*Brown* difficulties. As Harlan saw it, "the essential reason for the remand procedure chosen in *Brown II* was the Court's anticipation of difficulties of the sort we deal with in these cases."

Another Harlan suggestion that Burger accepted was the elimination of the italicized portion of the sentence "District courts must weigh the soundness of any transportation plan in light of what is said in subdivisions (1), (2), and (3) above, *each of which is in part a limitation on the extent of bus transportation of students.*" The change suggested, wrote Harlan, "is intended to eliminate any implication of a presumption against 'bussing' as such."

In addition, Harlan recommended a change that is of interest because of the continuing effort to secure congressional restrictions upon the remedial power of the federal courts—for example, by laws prohibiting them from issuing school busing orders. The Burger third draft began its discussion of the effect of the Civil Rights Act of 1964 by stating, "Had Congress chosen to limit the permissible remedies for a violation of the Fourteenth Amendment in school desegregation cases, the situation might be quite different. It is clear, however, that Title III of the Civil Rights Act of 1964 did not do that." Harlan suggested substitution of the following: "It is clear that Title III of the Civil Rights Act of 1964 was not intended to limit the permissible remedies for a violation of the Fourteenth Amendment in school desegregation cases."

The Justice explained the substitution this way: "For me the problem of the Congress' power to limit the remedial discretion of the federal courts once a substantive *constitutional* violation is shown is a very difficult one. I do not think we should give any indication of an affirmative view on the validity of such legislation."

Harlan also recommended that the Chief Justice omit the two sentences at the beginning of his Part VI, which referred to the court of appeals' "reasonableness" test and the *Green* test. "It would, I think," Harlan wrote, "be better to avoid any explicit reference to the Fourth Circuit's 'reasonableness' formula than to suggest that there is no real difference in emphasis between our opinion in *Green* and this case on the one hand and the Fourth Circuit's opinion in this case on the other hand."

Next to this part of his copy of his letter, Harlan wrote, "Reject." The offending sentences appeared in the Chief Justice's next draft and remain part of the final *Swann* opinion.

Harlan Redraft

Even more important than the individual suggestions contained in the Harlan letter was an attachment which the Justice entitled "Proposed Reorganization of Part V(1) of Draft Opinion." This was essentially a redraft of the Chief Justice's section on the first question dealt with in Part V of his third draft, headed "Racial Balances or Racial Quotas." This portion of the Burger draft had contained some of the passages to which the Justices who favored affirmance of Judge McMillan most objected.

Harlan's redraft was presented to the Chief Justice as primarily a rearrangement of the Burger version: "I have taken the liberty of suggesting a reorganization of these five pages." Using the Chief Justice's language as much as possible, Harlan had changed the order in which the different parts of the Burger section had been presented. In the process, he had eliminated the Burger passages that had been most offensive to some of the Justices.

Most important, the Harlan redraft omitted the Parker-like declaration in the Burger draft: "The Constitution, of course, does not command integration; it forbids segregation." Also eliminated was the statement that the McMillan opinion "contains intimations that the 'norm' is a fixed mathematical balance reflecting the pupil constituency of the system" and the similar statement "Some of the language of the District Court can indeed be read as intimating that racial balance is a constitutional requirement," as well as the assertion that a federal court was without power to require that any particular racial balance be maintained.

In his letter to the Chief Justice, Harlan gave the assurance that the "substantive points" made in the omitted sentences "are also made in the proposed revision." In actuality, those points were not contained in the Harlan redraft. They thus disappeared from the *Swann* opinion, making it easier for those who had opposed the Burger drafts now to consider joining the Chief Justice.

The opposition to the Burger drafts had by this time substantially weakened. When the Chief Justice circulated his first draft, the others decided that they could not go along with him—that (in the phrase of one Justice) "this simply won't wash." Then the Chief Justice rewrote his opinion and it, too, was unacceptable to his colleagues. In fact, as we saw, the second draft was as unsatisfactory as the first in its tone and its grudging attitude toward Judge McMillan. Despite the statement in the Chief Justice's covering letter that "the enclosed draft is revised throughout," the second draft was essentially similar to the earlier version.

At that point, the other Justices remained fixed in their determination not to have the Burger draft come down as the opinion of the Court. As the Justice already quoted told me, "The Chief Justice was determined to carry the day and we were equally determined that he wouldn't—and we had the votes." The supporters of Judge McMillan now focused on the Stewart draft as an alternative and prepared to work for its acceptance in place of the Chief Justice's version. But then came the Burger turnabout, which substantially changed the situation.

In his third draft, the Chief Justice had corrected much of the negative tone of his earlier versions and, most important, had ended with a clear affirmance of Judge McMillan, including the busing requirements ordered by him. The Burger change in position took most of the fire out of the opposition to his drafts. The McMillan supporters had, after all, secured what they most wanted: affirmance of the district judge and the signal that that would send to the lower courts. To most of them, this was far more important than continuing to refine the language used by the Chief Justice. This was particularly true after Justice Harlan had circulated his redraft. If the Harlan suggestions were accepted, the most offensive passages would be eliminated from the Burger opinion. Justices Harlan, Stewart, and Douglas had indicated that they might now be willing to join the opinion, particularly if some further refinements were made.

The one McMillan supporter who still pressed for more substantial changes in the Burger draft was Justice Brennan. His emphasis on language had become the hallmark of his judicial work and he continued to urge that *how* the *Swann* opinion said what it did was as important as *what* it said. But the others were tiring of the battle, which some of them thought was turning into a semantic conflict. The Harlan, Stewart, and Douglas letters to the Chief Justice had shown that they were ready to abandon the Brennan effort to secure further substantial changes. Brennan had lunch with Justice White to discuss the problem. Pressed by Brennan, White agreed to see the Chief Justice and ask him to withdraw his opinion in favor of the Stewart draft. When he saw Burger, however, White ended by telling the Chief Justice that he could also probably join his opinion.

10

Fourth and Fifth Drafts

Chief Justice Burger had not been able to obtain acceptance of his drafts refusing to affirm Judge McMillan. His attempt to control the drafting process and secure a *Swann* opinion reflecting his own restricted view of judicial remedial power in school desegregation cases had been rebuffed by the Justices who supported the district court's far-reaching orders. But the Chief Justice had now substantially yielded to the wishes of the others. His third draft had given McMillan's supporters what they most wanted: a clear affirmance of the district judge. In addition, the new draft had specifically approved McMillan's busing order and omitted some of the language that the others found objectionable. It is true that it still contained passages that were unacceptable to some, notably the statement that the Constitution only forbade segregation and did not command integration. This Parker-like pastiche was particularly offensive to Justice Brennan, who thought that his own *Green*[1] opinion had relegated the Parker principle[2] to a deserved oblivion.

With a little redrafting, however, the Burger third draft could be rendered acceptable to those who had been dissatisfied with the prior versions. As already seen, the Chief Justice's section on racial balance had been rewritten by Justice Harlan. The Harlan redraft kept most of Burger's language, but omitted the sentences that had offended the others, notably the aforementioned assertion on the Constitution and integration. If the Chief Justice accepted the Harlan redraft and a similar pruning job was done on other portions of his opinion, the result could be joined by the opponents of the earlier drafts, with the possible exception of Justice Brennan. And even if he remained not wholly satisfied with the opinion, he would feel strong pressure not to stand alone and break the tradition of unanimity in school desegregation cases.

Fourth Draft

I have been told that the major role in the final redrafting of the *Swann* opinion (apart from the Harlan redraft of the racial balance section) was played by Justice Stewart. It has, however, been impossible to document this assertion, which was made to me on a confidential basis. What can be documented is the fact that on March 16, the day before the next scheduled conference on the case, the Chief Justice circulated his fourth *Swann* draft.

The draft was accompanied by a *Memorandum to the Conference:*

> Since my last preceding draft I have consolidated most of the specific suggestions submitted by various members of the Court. The enclosed draft is therefore in a sense 'provisional,' *i.e.,* I have incorporated suggestions with a view to accommodating views that might otherwise have been expressed separately. If we have now achieved that objective, this draft will supplant the prior drafts.[3]

The fourth Burger draft consisted of 36 typewritten pages. Its introductory statement and Part I (containing the facts and a history of the case) were the same as those in the third draft. Part II (discussing the *Brown* decisions[4]) was also based on the version contained in the prior draft. There were, however, some changes which should be noted. The new draft followed the Harlan suggestion and omitted the statement that the "detail and nature" of the dilatory tactics used in opposition to *Brown* "are not now relevant." Also eliminated was the phrase "no one could foresee in 1954" in referring to the "practical difficulties" encountered after *Brown*. In addition, there was a direct reference to southern resistance to desegregation: "The failure of local authorities to meet their constitutional obligations aggravated the massive problem of converting from the state-enforced discrimination of racially separate school systems."

Part III of the Burger fourth draft began, "The objective today remains to eliminate from the public schools all vestiges of state-imposed segregation. That was the evil struck down by *Brown I* as contrary to the Equal Protection guarantees of the Constitution. That was the violation sought to be corrected by the remedial measures of *Brown II*." Though the first sentence was slightly changed, this passage was essentially the same as it had been in the prior draft, with its implication that the judicial goal was only the elimination of "state-imposed segregation."

The negative aspect of this Burger statement was, however, now modified by the addition of a quotation from *Green* on the school authorities' "affirmative duty to take whatever steps might be necessary to convert to a unitary system in which racial discrimination would be eliminated root and branch," as well as the flat assertion that "If school authorities fail in their affirmative obligations under these holdings, judicial authority may be invoked." Following this was a stronger statement of the broad remedial power of the district court.

Though the Chief Justice still repeated "that judicial powers are not co-extensive with those of school authorities," the later portions of the relevant passage were changed. The third draft had referred to the establishment of "a fixed ratio of Negro to white students reflecting the proportion for the district as a whole." It had asserted, "To do this as an educational policy is within the broad discretionary powers of school authorities but is beyond judicial authority." The fourth draft changed this to: "absent a finding of a constitutional violation, however, that would not be within the authority of a federal court." The new draft then noted that a federal court may intervene "if . . . a constitutional violation is shown." In this case, "the nature of the violation determines the scope of the remedy. In default by the school authorities of their obligation to proffer acceptable remedies, a district court has broad power to fashion a remedy that will assure a unitary system."

This was a complete change from the restricted conception of judicial authority contained in the Burger first draft. Now it was made clear that once a constitutional violation was found, the remedial authority of the district court was as broad as necessary to correct the violation, including any measures needed to "assure a unitary system."

The Chief Justice also revised the beginning of his discussion of the 1964 Civil Rights Act in accordance with the suggestion in the March 11 Harlan letter.[5] The two sentences intimating that Congress might limit school desegregation remedies were eliminated and the discussion now began with the Harlan statement that the civil rights statute was not intended to limit such remedies. Incidentally, the Burger draft now referred correctly to Title IV of the Civil Rights Act. (The second and third drafts had incorrectly referred to Title III, as had the Douglas memorandum of October 17, 1970, upon which their discussion had been based.)

Part IV of the fourth draft was essentially the same as the version in the prior draft. The one change of consequence came in a discussion of school construction to perpetuate dual schools. Both drafts noted, "Such a policy . . . may well promote segregated residential patterns which, when combined with 'neighborhood zoning,' further lock the school system into the mold of separation of the races." The fourth draft added, "Upon a proper showing a district court may consider this in fashioning a remedy." In addition, the draft now stated that it was the district court's duty to see that "future school construction and abandonment is not used and does not serve to perpetuate *or reestablish* the dual system"—adding the italicized words to the third draft's version.

Part V of the fourth draft began, as had the third, with the statement of the four questions related to the "central issue . . . that of student assignment." The first question had, however, been rephrased: "to what extent racial balance or racial quotas may be used as an implement in a remedial order to correct a previously segregated system."

The discussion of the first question, that of racial balance, now contained the version suggested in Justice Harlan's redraft. As noted in the last chapter, this meant the elimination of most of the Burger passages that had been most offensive to the Justices who strongly supported Judge McMillan. Most important of all were the omissions mentioned in the last chapter: the Parker-like proposition on the Constitution not commanding integration, but only forbidding segregation; the statements that McMillan had intimated that the "norm" was a fixed racial balance and that such a balance was required by the Constitution; and the assertion that a federal court might not require any particular racial balance.

Instead, the new draft contained a much milder version: "The Constitutional command is to desegregate schools, but this does not mean that every school in every community must reflect the racial composition of the school system as a whole. There is no constitutional requirement that any particular racial balance or ratio be permanently maintained." This language was not taken from the Harlan redraft, yet it was plainly an improvement over the previous Burger draft, since the statements referred to in the preceding paragraph had been eliminated.

The notion that the Constitution does not require a rule of racial balance had been taken from Justice Stewart's draft dissent. In addition, Part V(1) of the new Burger draft ended by relying on another portion of the Stewart draft, where Stewart had written, referring to Judge McMillan, "he took the 71/29 ratio of white to Negro students in the schools as no more than a starting point for the process of planning, rather than as an inflexible requirement. From that starting point he proceeded to a final decree which was fully within his discretionary power to frame, as an equitable remedy in the particular circumstances that confronted him."

The version in the fourth Burger draft read, "the use made of mathematical ratios was no more than a starting point in the process of shaping a remedy, rather than an inflexible requirement. From that starting point the District Court proceeded to frame a decree that was within its discretionary powers, an equitable remedy for the particular circumstances." This statement was followed by specific recognition of "the need for limited use of the mathematical ratios in the transition from a dual to a unitary system."

The new Burger's draft's discussion of the second question, that of one-race schools, still began by referring to "the most valiant efforts" of school authorities. The remainder of this section was unchanged from the third draft, except for the addition of a paragraph, as Justice Harlan had suggested in his March 11 letter. The new paragraph, however, combined language taken from both Justices Harlan and Stewart. It began:

> In light of the above, it should be clear that the existence of some small number of all-Negro, or virtually all-Negro, schools within a district is not in and of itself the mark of a system which still practices segregation by law. The district judge or school authorities working to achieve the greatest possible degree of actual desegregation will necessarily be vitally concerned with the elimination of all-Negro schools. No per se rule can adequately embrace all the difficulties of reconciling the competing interests involved.

This language was taken almost verbatim from the Stewart draft. The remainder of the added paragraph stated that the need for sufficiently specific remedial criteria warranted a presumption against schools substantially disproportionate in their racial composition. In systems with a past history of segregation, courts must give close scrutiny and school authorities have the burden of showing that school assignments are nondiscriminatory. This added language was based on the suggested new paragraph contained in the Harlan March 11 letter.

Part V(3) of the Burger fourth draft, dealing with remedial altering of attendance zones, was essentially unchanged from the third draft. The Chief Justice did, it is true, change the passage to which the Harlan letter had taken strong exception, on the objective being the achievement of student assignment and distribution that would have existed had there not been previous discrimination. The new draft did not omit this statement, as Justice Harlan had urged. Instead, it was altered to read, "The objective is to eliminate the dual school system, i.e., to achieve as nearly as possible that distribution of students and those patterns of assignments that would have normally existed had there been no state enforced segregation." This was, of course, only a change in phrasing and scarcely met the Harlan objection.

Part V(4), on busing, was also virtually unchanged, though the new draft did omit the statement "each of which is in part a limitation on the extent of bus transportation," after saying that district courts must weigh the soundness of busing plans in light of what was said in the prior portions of Part V. This was in accordance with another suggestion in Justice Harlan's letter, which explained that "the change is intended to eliminate any implication of a presumption against 'bussing' as such."

The fourth draft's concluding Part VI rejected the Harlan letter's recommendation to omit the first two sentences referring to the court of appeals' "reasonableness" test and the *Green* test. The new draft did, however, leave out the statement that the Court was unable to conclude that the district court order "is not reasonable, feasible and workable." Instead, it asserted, "Substance, not semantics, must govern, and we have sought to suggest the nature of limitations without frustrating the appropriate scope of equity."

The rest of Part VI remained unaltered from the third draft. Once again the draft concluded: "The judgment of the Court of Appeals is affirmed as to those parts in which it affirmed the judgment of the District court. The order of the District Court dated August 7, 1970 is also affirmed."

Douglas Note and Dissent

The Harlan redraft of Part V(2) had contained changes which, the Justice's letter explained, were made "in light of Brother Douglas' wish to reserve the other elements of state action to a 'non-dual system' case." The Chief Justice followed the Harlan suggestion, eliminating the sentence "Discrimination in other areas of life must be dealt with by other remedial mechanisms based on constitutional or statutory guarantees." He also omitted the italicized portion from the statement that the Court's objective "cannot embrace all the problems of racial prejudice *in residential patterns, employment practices, location of public housing, or other factors beyond the jurisdiction of school authorities,* even when those patterns contribute to disproportionate racial concentrations in some schools."

The change was, however, not enough to appease Justice Douglas. On March 16, he circulated a "Dear Chief" note stating that he was now "unable to join your opinion." According to Douglas, "The case is rather a simple one, since it deals with a situation where there was a dual school system. Hence we need not get into other types of *de jure* segregation."

"Your opinion," the Justice further objected, "decides that those other kinds are not relevant to any problem the federal court may have in working out a school integration program." That issue, Douglas noted, had never been argued and "I think that the decision of that question in the *Swann* case is needless."

Justice Douglas wrote that he had hoped to join the Chief Justice, but now had decided, "I . . . will write separately." The promised opinion was circulated by the Justice on March 19, the day after the scheduled conference on the case. Headed simply "Mr. Justice Douglas," it contained seventeen printed pages and began, "This case is a simple one that should be quickly dispatched." The remainder of the Douglas draft was not materially different from the draft opinion circulated by the Justice on January 13.

Further Brennan Objections

The Douglas objection could readily be dealt with by the Chief Justice. More difficult were the continuing objections raised by Justice Brennan. He sent the Chief Justice a nine-page letter on March 17, the day before the conference, which made sixteen suggestions. The most important of these will be summarized.

The first Brennan suggestion was that the opinion eliminate all indications that the case presented "unresolved issues" or "many unanticipated difficulties." The Justice stressed "that the difficulties in which these Schools Boards find themselves are primarily the consequence of their own delaying tactics in complying with *Brown I*." He thought that this should "be made crystal clear" in the opinion.

Brennan stated that he still thought "that your treatment of the distinction between School Board powers and judicial powers reads to me as a retreat from what we said in *Green*." The Justice suggested "that this gloss can be removed" if the Burger statement "that judicial powers are not co-extensive with those of school authorities" was revised to read, "In seeking at this stage to define even in broad and general terms how far this remedial power extends, it is important to remember that judicial powers may be exercised only on the basis of a constitutional violation."

Brennan also wrote, "I agree with Bill Douglas that the implication remains that other types of *de jure* segregation are not relevant to any problem the federal court may have in working out a school desegragation program. I think we should say flatly that that question is not being decided in this case."

In particular, the Justice urged a change in the sentence in the Burger draft that read, "It would not serve the important objective of *Brown I* to seek to use it for purposes beyond schools; desegregation of schools will ultimately have impact on other forms of discrimination."

According to Brennan, "the sentence seems to say that *Brown* itself has no application where schools are not involved. This is clearly not so, as our *per curiams* dealing with other forms of state-imposed segregation following *Brown* make clear."

The Justice suggested that the offending sentence should read, "It would not serve the important objectives of *Brown I* to use school desegregation cases for purposes beyond their scope, although desegregation of schools should ultimately have an impact on other forms of discrimination."

Brennan also urged that the Chief Justice's continuing reference to the "most valiant efforts" of the schools boards be deleted, as well as a statement in Part V(2) on one-race schools, which, in Brennan's characterization, "asks the district courts to conjure up the situation which 'would have existed independent of any previous discrimination.' "

The Justice then suggested a rewriting of the Burger passage on majority-to-minority transfer provisions that had said that such provisions "will tend to relieve particular hardship cases for those who find the posture of being part of a school racial majority a 'badge of inferiority.' " To Brennan, "the sentence for me is uncomfortably reminiscent of *Plessy* v. *Ferguson*,"[6] the 1896 case that had originally upheld segregation. Brennan pointed out that the Burger phrase had come from the statement there that read, "We consider the underlying fallacy of the plaintiff's argument to consist in the assumption that the enforced separation of the two races stamps the colored race with a badge of inferiority. If this be so, it is not by reason of anything found in the act, but solely because the colored race chooses to put that construction upon it."[7] The Justice suggested a revised wording for the majority-to-minority transfer provision which the Chief Justice used in his next draft.

Brennan also strongly agreed with Justice Harlan on the deletion of the Burger test: "The objective is to eliminate the dual school system, i.e., to achieve as nearly as possible that distribution of students and those patterns of assignments that would have normally existed had there been no state enforced segregation." Though, as seen, the Chief Justice had not followed the Harlan suggestion in his fourth draft, after the Brennan objection he did omit the sentence in his fifth draft.

The Chief Justice was to reject another Brennan suggestion, that he delete the statement in his busing section that maximum travel time would vary most with the age of students. This, according to the Justice, "means . . . that young children may not be transported as far as older ones." Brennan wrote, "I am unable to find any basis in the record for your underlying assumption that young children are less able to adapt to long trips to school than are older ones. This may or may not be true. But I hardly think we should decide the question on the basis of our own, unsubstantiated opinions."

The Justice also objected to the end of the Burger busing section, which referred to "the competing values of the need to grant a remedy for violation of constitutional rights, on one hand, and the well-being of children, on the other." Brennan further asserted, "Your formulation of the problem poses a conflict between an abstract, albeit important value— constitutional legality—on the one hand, 'and the well-being of children, on the other.' But the whole thrust of *Brown I* is that state-imposed segregation of schools is not a mere abstract evil." From this point of view, the Justice declared, "to pose the problem as one of competition between abstract values on the one hand and the welfare of children on the other is to misunderstand the evil of segregation. The welfare of Negro children is inextricably bound up with elimination of unconstitutional discrimination against them. If we ever permit the District Courts to neglect this basic truth, we can never hope for full implementation of *Brown I.*"

In addition, the Justice agreed with Justice Harlan that the Chief Justice should delete the reference to the court of appeals "reasonableness" test and the *Green* test that began the last section of his fourth draft. As noted in the last chapter, this recommendation was rejected.

Justice Brennan ended his letter, "I repeat the thought in the last sentence of my March 8 comments: Potter's draft so fully meets the objections I've made, that I'm prepared to join him, and can't join any opinion which does not follow his lead." Once again the Justice had threatened that he was ready to join Justice Stewart's draft. Unless the Chief Justice further revised his opinion, the tradition of unanimity in school segregation cases could be broken.

Work Draft and Conference

Justice Brennan's threat not to join the *Swann* opinion was, by this point, one that could readily be countered by the Chief Justice. The Brennan letter had said that the Justice was prepared to join "Potter's draft," but Justice Stewart had never circulated his opinion and was now all but ready to agree to the latest Burger draft. Like most of the others, Stewart had grown weary of the struggle and concluded that the decision for affirmance was more important than further fine-tuning of the Burger language. This was true even of Justice Douglas, who, despite his note and circulated draft, would join if the Chief Justice expressly stated that the Court was not deciding on any other types of segregation. Despite his threat, Brennan would scarcely stand alone in refusing to join the *Swann* opinion, particularly if the next Burger draft made the gesture of adopting at least some of the suggestions in the Justice's March 17 letter.

At this time, however, Justice Brennan—alone or not—had not yet given up his effort to refine further the Burger draft. He had sent his list of suggestions to the Chief Justice the day before the scheduled conference on the case and he arrived at the meeting the next morning prepared to do battle over them during a long day's session. To aid the discussion, the Justice had prepared what he called a "work draft."[8] This was the Brennan copy of the Chief Justice's fourth draft, with the suggestions of the different Justices worked into their proper places. All the necessary changes were made in Brennan's handwriting.

The March 18 conference (the first held on the case since December 3, 1970) finally arrived, but the long session expected by some of the Justices did not take place. "The conference," according to one Justice, "was over in ten minutes. The C.J. waltzed in and announced that he had read the suggestions of his colleagues and that he was content to make all of them in a new circulation." When Justice Brennan produced his work draft and prepared to discuss it, the Chief Justice said that he would take the copy in order to make the necessary changes.

After the session, Justice Marshall expressed his elation to Justice Brennan about the turn of events. The latter was highly dubious, noting that the Chief Justice had managed to prevent the Justices from discussing the matter when all were present at the same time and place.

The Brennan skepticism appeared justified when, the day after the conference, he received a "Dear Bill" letter from the Chief Justice, together with "your 'work draft' of my March 16 circulation." The Burger letter stated that the Chief Justice had reviewed "your suggestions closely," but "it becomes clear to me that I will not be able to accept all of them. Some are just the difference between the way in which two people express the same ideas but others seem to go beyond what at least five are prepared to accept."

The Justice thought that, despite the Chief Justice's assertion, it was not at all clear that he had five who were prepared to join his opinion. Yet the Burger letter did indicate that he was willing to meet Brennan at least part of the way: "I can and will accept most of the 'tonal' changes and the abstention from the suggestion of a holding on what is *de jure* or *de facto* segregation—I hope without using those increasingly misused terms."

The last point was, of course, intended to cover the Douglas objection. If that could be done and "the 'tonal' changes" could eliminate most of the passages which some of the Justices still found objectionable, the Court would be very close to having an opinion with which all could agree.

Fifth Draft

On March 22, the Chief Justice circulated his fifth *Swann* draft. It was the first printed draft of the *Swann* opinion and contained 28 printed pages. It was accompanied by a *Memorandum to the Conference* which stated, "In the hope of achieving a unanimous opinion I have made substantial changes and now circulate what I trust is the final draft of the *Swann* opinion. I have adopted the express reservation of the broader '*de jure*' question suggested by Justice Douglas and others."

The Burger memo noted changes made in the standards on one-race schools and the omission of references to the resegregation problem. These will be dealt with in the summary of the draft that follows. "For the rest," the memo went on, "I have adopted most of Justice Brennan's suggestions, or at least attempted to modify them to an acceptable compromise."

The beginning of the new draft bore out the Chief Justice on this point. The first sentence omitted the indication that the case presented "unresolved issues," to which Justice Brennan had taken exception—as well as the statement that "this Court has not resolved the specific issues raised in" the case. Brennan had objected to this in his memoranda of December 30, 1970, and March 8, 1971, since he thought that *Green*[9] had all but settled the law on the matter.

The Chief Justice also followed the Brennan work draft by transferring to the second paragraph of the new draft the material (derived in part from the Stewart draft dissent) that had appeared in Part II of the fourth draft. The transferred passage referred to the need to define in more precise terms the scope of the duty to implement *Brown* and "the mandate to eliminate dual systems and to establish unitary systems at once," in addition to the need for guidelines that would take into account the experience of the lower courts. This passage provided a missing focus for the beginning of the opinion and indicated that the Court was going to do what Justice Harlan had urged at the very beginning of the Justices' deliberative process: formulate guidelines for future cases.

Part I of the Burger fifth draft was taken almost verbatim from the fourth draft—with one significant exception. All the previous drafts had contained a footnote 3, after the statement of facts, referring to the district court's appointment of Dr. Finger as an expert to prepare a desegregation plan:

Although Dr. Finger had previously appeared as a witness for one of the parties, the Court of Appeals found that this posture did not cause him to be faithless to the trust the court imposed on him and held the error of his dual role, if any, harmless. We adopt that view but with it a caveat as to the future. Expert witnesses with an interest in sustaining their own plans and positions are placed in an awkward position at best.

In his work draft, Justice Brennan had crossed out this footnote and written next to it, "I would omit this." The Chief Justice followed this suggestion and the footnote does not appear in the fifth draft or final *Swann* opinion. The omission was fortunate. It would have been regrettable if an opinion upholding Dr. Finger's plan had cast any doubt upon his impartiality as an expert, especially in view of the virtual impossibility of securing any other expert which (as we saw in Chapter 1) had faced Judge McMillan in the district court.

Part II of the Burger fifth draft was also basically similar to the version in the previous draft. As already noted, a passage from this part was moved up to the beginning of the opinion. In addition, the Chief Justice finally yielded to Justice Brennan's continuing objection to the reference to the "many unanticipated difficulties" encountered since *Brown II.* The new draft eliminated "unanticipated." Burger, however, refused to follow the Brennan suggestion that he delete the sentence "Nothing in our national experience *prior to 1955* prepared anyone for dealing with changes and adjustments of the magnitude and complexity encountered *since then,*" though he did add the words in italics and strengthen the next sentence so that it read, "Deliberate resistance of some to the Court's mandates has impeded the good-faith efforts of others to bring school systems into compliance." These changes went at least part of the way to meet the Brennan objections.

In Part III, the fifth Burger draft at last gave way on the statement that others had found offensive, "that judicial powers are not co-extensive with those of school authorities." In its place, the draft adopted the sentence suggested in the Brennan March 17 letter: "In seeking to define even in broad and general terms how far this remedial power extends it is important to remember that judicial powers may be exercised only on the basis of a constitutional violation." This finally removed the inference that the remedial power of the federal courts was somehow more restricted here than in other cases.

The remainder of Part III was unchanged from the prior draft. The same was true of Part IV, where there was only a minor stylistic change, suggested by Justice Brennan.

Part V of the fifth draft began with the four questions related to student assignment, as they had been phrased in the prior draft. The discussion of the first question, that of racial balance, contained several significant changes. As indicated in his covering memorandum, the Chief Justice had "adopted the express reservation of the broader 'de jure' question suggested by Justice Douglas and others." He did this, first of all, by including the sentence suggested by Justice Brennan on the objective of *Brown I* not being served by using "school desegregation cases for purposes beyond their scope, although desegregation of schools ultimately will have impact on other forms of discrimination." This was followed by an express disclaimer of intent to decide on any other types of *de jure* segregation: "We do not decide whether other types of state-imposed segregation may be relevant to the problem of working out a school desegregation program. This case does not present that question and we therefore do not decide it." These two sentences had been added in Justice Brennan's work draft.

In addition, the new draft omitted the sentence "There is no constitutional requirement that any particular racial balance or ratio be permanently maintained"—as Justice Brennan had suggested. Another Brennan suggestion—originally made in Justice Harlan's redraft of Part V(1)—was also adopted: to end the discussion of racial balances with the statement "In sum, the limited use made of mathematical ratios was within the equitable remedial discretion of the District Court."

The fourth Burger draft discussion of racial balances had ended with a paragraph on the need for future action to deal with the "resegregation" problem. The passage was deleted from the fifth draft to meet another Brennan objection. As the Chief Justice explained it in his March 22 covering memorandum, "in the interests of achieving unanimity I have omitted all the references to the 'resegregation' problem even though I believe they had the support of a majority."

The fifth draft's section on one-race schools finally omitted the conciliatory reference to the "most valiant efforts of school authorities," which had so offended some of the Justices. In his March 17 letter, Justice Brennan suggested that the sentence containing the offending phrase be rewritten to read, "In some circumstances certain schools may remain all or largely of one race until new schools can be provided or neighborhood patterns change." The Burger fifth draft adopted this suggestion. It also omitted the next sentence in the fourth draft: "We find nothing in the Constitution read in its broadest implications that requires invariably the elimination of schools all or predominately of one race in a district of mixed population so long as school assignments are not part of state-enforced segregation." The omission had been urged in the Brennan letter on the ground that "the tone of the sentence, although its substance is correct, is for me unacceptably negative."

The previous draft had spoken of the burden on school boards "of showing that such school assignments are truly non-discriminatory. The burden of the school authorities will then be to satisfy the court that the imbalance found would have existed independent of any previous discrimination." Justice Brennan had objected to this as requiring "the district courts to conjure up the situation which 'would have existed independent of any previous discrimination.'" The Burger fifth draft changed the passage to read, "The burden upon the school authorities will be to satisfy the court that their racial composition is not the result of present or past discriminatory action on their part."

Justices Harlan and Brennan had objected strongly to some of the language in Part V(3), on the remedial altering of attendance zones—notably to the statement that the objective in such a case was to achieve the student distribution and assignment patterns that would have existed had there been no state-enforced segregation. The Chief Justice had refused to delete this language in the prior draft. But now it disappeared from the fifth draft. The rest of Part V(3) remained unchanged from the fourth draft.

Part V(4) of the Burger fifth draft, on busing, was essentially the same as that section in the previous draft. The Chief Justice rejected the Brennan suggestion that he delete the sentence indicating that time limits on travel would vary most with the age of students. The new draft did, however, follow Brennan's urging and eliminate the reference to "the competing values of the need to grant a remedy for violation of constitutional rights, on one hand, and the well-being of children, on the other."

The concluding section of the Burger fifth draft started with the paragraph that had begun Part VI in the fourth draft. The Chief Justice had again declined to accept the Harlan suggestion (repeated in Justice Brennan's May 17 letter) that he omit the references to the court of appeals "reasonableness" test and the *Green* test. As explained in the Burger March 22 covering memorandum, "I have not changed the first full paragraph of Part VI since I am in receipt of conflicting requests . . . the particular words are not crucial and is no more than a general caveat at most."

Two other Brennan suggestions were also rejected. The first was that the Chief Justice add back the sentence omitted from the fourth draft: "On the facts of this case we are unable to conclude that the order of the district court is not reasonable, feasible and workable." Though the fifth draft did not follow this Brennan suggestion, the sentence was to be put back in the Chief Justice's sixth and final draft.

Justice Brennan had also objected to the fourth draft's "resegregation" paragraph near the end, which, as the Justice put it in his March 17 letter, "strongly implies that the federal courts should get out of school cases soon, indeed 'in the near future.' " In the Brennan work draft, this paragraph had been crossed out. The Chief Justice retained the paragraph in his fifth draft, though he did delete the phrase "in the near future" and the sentence "It should be clear that desegregation decrees are not an appropriate instrument for judicial monitoring of shifting residential patterns unrelated to discriminatory school segregation."

The fifth Burger draft ended much as did the two prior drafts (with the addition of the italicized introductory phrase and the italicized traditional concluding formula):

> *For the reasons herein set forth,* the judgment of the Court of Appeals is affirmed as [to] those parts in which it affirmed the judgment of the District Court. The order of the District Court dated August 7, 1970, is also affirmed.
>
> *It is so ordered.*

11

Threatened Black Dissent and Final Draft

"Your last circulation is mighty close," wrote Justice Marshall in a "Dear Chief" letter the day after the fifth *Swann* draft was circulated.[1] The Marshall statement represented the consensus of the Justices after they had read the latest Burger draft. They realized that, even in its latest version, the *Swann* draft was scarcely a sparkling opinion; but they also recognized that the fifth draft was much improved over the earlier versions. What Justice Brennan had termed the drafts' "negative tone" was now all but eliminated and almost all the language that had been offensive to the supporters of Judge McMillan had been either removed or toned down. The drafting contributions of Justices Harlan, Stewart, and Brennan had all helped to make the opinion more acceptable. Most important, the draft now provided for a firm affirmance of Judge McMillan and express approval of what he had done—particularly of the busing requirements he had imposed.

All in all, to McMillan's supporters, the Chief Justice's latest draft represented a quantum leap forward from his utterly unacceptable first draft. Certainly, the Court was now "mighty close" to a *Swann* opinion which the Justices could finally join.

Final Brennan Effort

On March 23, the day after he had received the Burger fifth draft, Justice Brennan indicated that he still had difficulties with the *Swann* opinion. He sent a final set of suggestions to the Chief Justice. The Brennan letter stated that the Court was now very near to a unanimous opinion. "In that spirit," the Justice wrote, "I can accept almost all of the present circulation. I am still troubled, however, about four matters and offer the following suggestions for your consideration."

The first suggestion related to the reservation of the issues involved in other types of de jure segregation. Brennan thought that "the wrong question has inadvertently been reserved" in Part V(1) of the fifth draft. The draft had stated, "We do not decide whether other types of state-imposed segregation may be relevant to the problem of working out a school desegregation program. This case does not present that question and we therefore do not decide it."

According to Brennan, the lower courts had found that such other state-imposed discrimination existed: "In order to affirm Judge McMillan, I think we necessarily approve taking account of these findings in the context of a remedial decree, once a constitutional violation has been found on the basis of the school board's action itself." Because of this, the Justice suggested, "The question not presented and on which decision should be reserved, I believe, ought to be phrased as follows: 'We do not decide, however, whether a constitutional violation is made out by a showing that other types of state-imposed segregation have produced a segregated school system, even though no discrimination is shown on the part of school authorities.' "

The Justice's second suggestion related to the burden imposed on school districts by the existence of one-race schools. Brennan suggested adding language contained in Justice Harlan's March 11 letter to the Chief Justice, so that the key sentence in the "burden" paragraph would read, "Where the school authority's proposed plan for conversion from a dual to a unitary system contemplates the continued existence of some schools that are all or predominantly of one race, they have the burden of showing that such school assignments *truly reflect compelling countervailing considerations of educational policy or pupil welfare unrelated to community resistance to desegregation*"—with the added language in italics.

The third Brennan suggestion related to the discussion in Part V(3) of the Burger draft of "racially neutral assignment plans," which "I strongly fear . . . could be read to mean that a plan which is truly neutral on its face is sufficient." The Justice suggested revising the paragraph in point to read:

No fixed or even substantially fixed guidelines can be established as to how far a court can go, but it must be recognized that there are limits. The objective is to *dismantle* the dual school system, and in achieving this objective an assignment plan is not acceptable simply because *it is* neutral. Such a plan may fail to counteract the continuing effects of past school segregation resulting, for example, from discriminatory location of school sites or distortion of school size in order to achieve or maintain an artificial racial separation. When school authorities present a district court with a "loaded game board," affirmative action in the form of remedial altering of attendance zones is proper to achieve truly non-discriminatory assignments.

The Justice's last points related to Part VI. "In view of your present feeling . . . that the present formulation should remain," Brennan no longer asked for the deletion of the first two sentences on the court of appeals "reasonableness" test and the *Green* test.[2] Instead, he asked for the addition after them of the sentence that the Chief Justice had taken out earlier from his draft: "on the facts of this case, we are unable to conclude that the order of the district court is not reasonable, feasible and workable."

Brennan also suggested that the following language from Justice Stewart's draft be inserted in the paragraph on future population changes near the end of the Chief Justice's draft:

It is, of course, the responsibility of the district court to devise decrees which minimize in so much as foreseeable the probability of resegregation. The danger of resegregation may justify the requirement of a greater degree of actual desegregation of particular schools than would be necessary given a likelihood that populations will remain stable. Where a decree creates a disparity between schools or between areas which invites the migration of white parents from one school zone to another, it does not adequately perform its function of dismantling the dual system.

As Brennan saw it, "the reference to shifting demographic patterns will be misinterpreted without including the caveat from Potter's draft that desegregation decrees must take into account the possibility of resegregation."

Brennan Suggestions Endorsed

Soon after Justice Brennan had sent his suggestions to the Chief Justice, they were endorsed by four of the others. On March 23, Justices Marshall, Douglas, and Harlan wrote to the Chief Justice to tell him that they agreed with all of the Brennan proposals. The next day Justice Stewart sent a similar letter. The theme of all four letters was that stated in Justice Harlan's note: "I do think that the suggestions made in Bill Brennan's letter of today would improve and strengthen the opinion still more, and your adoption of all of them would be agreeable to me."

The Douglas letter also referred to the point on which he had consistently disagreed with the Burger drafts: the need for an express disclaimer of intent to decide other types of de jure segregation cases. "I guess," Douglas wrote, "there has been no meeting of the minds on my earlier suggestion, due doubtless to my informal wording." Douglas thought that the Brennan suggestion on the point "is the best solution." Harlan also wrote that Brennan's "suggested change in the 'reservation' respecting the broader state action question is a distinct improvement, and would be a desirable one to make."

The Marshall letter objected to the Burger draft's provision for a majority-to-minority transfer plan. The relevant paragraph, Marshall wrote, "gives me too much of a problem. It will inevitably result in the more affluent and educated Negro parents using the plan and leaving the poor Negroes stuck in the all-Negro school."

Marshall had come to the Brennan chambers to speak to the Justice about his misgivings on the matter. Finding Brennan absent, he spoke to one of the Justice's law clerks about his objection. Marshall also told the clerk that he disliked the use of "black" in preference to "Negro," since he was not "Black."

Though Justice Marshall was the only black on the Court and presumably was better informed about the practical impact of the transfer provision on the black community, his objection at this juncture came too late. The transfer provision had originally been proposed by Justice Stewart, accepted by the Chief Justice and the others, and refined by Justice Brennan. Now, after the provision had appeared in four drafts, they refused to accept an objection that should have been made months before. The Stewart March 24 letter to the Chief Justice took express exception to the Marshall point. "Although I understand that suggestion," Stewart wrote, "I cannot agree with it." As will be seen, however, the final *Swann* draft did make an attempt partly to meet the Marshall objection.

Black Objections

The Court was, in Justice Marshall's phrase, now "mighty close" to a consensus on the Burger fifth draft. True, Justice Brennan was still trying to refine the opinion and there were objections by Justices Douglas and Marshall to specific passages, but these were all on matters of detail, which would scarcely prevent agreement on the opinion's essentials. All that was necessary, it appeared, was a new draft that would make the minor changes needed so that all could join. As Justice Brennan had put it in his March 23 letter, "we are on the threshold of the unanimous opinion that all of us have been striving to achieve."

At this point, however, the consensus that had been so carefully constructed suddenly appeared in danger of collapse. Justice Black had expressed the strongest opposition to Judge McMillan's order at the first two *Swann* conferences. Since then, however, he had taken no direct part in the Court's decision process, though he had indicated his continuing objections to the Chief Justice and some of the others. Now, just when it appeared that the Justices were ready to agree on the *Swann* opinion, Black sent a March 25 letter to the Chief Justice threatening to dissent if changes were not made in the opinion. Worse still, the letter was accompanied by a draft dissent that would have destroyed the unanimity on the merits that had prevailed thus far in school segregation cases.

Justice Black had, of course, been the leader of the Court's liberal wing during his early years on the bench. He always thought that it was he who was primarily responsible for the judicial revolution that took place while Earl Warren was Chief Justice. Toward the end of the Warren tenure, however, as seen in Chapter 2, the Alabaman began to display a more conservative attitude than some of his colleagues. Black's fundamentalist approach to the Constitution did not permit him to adopt the expansive approach toward enforcement of individual rights followed by some of the Justices.

The more restrictive Black posture first manifested itself in cases involving civil rights protests. The Justice had, to be sure, been firm in voting for judicial enforcement of the Constitution's guaranty of equal protection. But he refused to go along when civil rights protesters attempted to secure vindication of these rights in ways he considered incompatible with the rule of law, upon which, he argued, all rights ultimately rested. The cases involving prosecution of sit-in demonstrators that came before the Court during the 1960s marked Black's constitutional divide. In the conflict of basic interests involved in them, the Justice came down on the side of property and the preservation of public order. During a conference on one of the sit-in cases, Black emotionally declared that he could not believe that his "Pappy," who ran a general store in Alabama, did not have the right to decide whom he would or would not serve.

Now the Justice had the same emotional reaction to judicial efforts to require people to have their children bused to schools out of their own neighborhoods. At the first *Swann* conference, on October 17, 1970, Black had forcefully asserted his view against Judge McMillan's busing order. In a statement already quoted, the Alabaman declared, "I have always had the idea that people arrange themselves often to be close to schools. I never thought it was for the courts to change the habits of the people in choosing where to live." At the same conference, Black indicated that he disagreed with any requirement of racial balance in schools. "The Constitution," he said, "doesn't require a particular proportion." During the next few months, while the *Swann* opinion was being worked out, the Justice remained fixed in his view. He expressed his disapproval of what was being done in the successive Burger drafts, particularly to the Chief Justice himself. Now Black decided that it was time to exact his due in the decision process. On March 25, as indicated, the Justice sent a "Dear Chief" letter which stated that, while he "had hoped to go along with your opinion," he found now "that there are statements adopted at the request of other Justices which I shall find it difficult, if not impossible, to accept."

The first statement to which Justice Black objected was the passage in the Burger discussion of the Civil Rights Act that said that the relevant section "was not enacted to limit but to expand and define the role of the Federal Government in the implementation of the *Brown I* decision."[3] Black asserted that this statement was not accurate; the act and its legislative history "appear to me to show that the purpose of the Congress was to deny the courts exactly what the complainants in this case are urging, namely, that pupils be transported to achieve a racial balance. Consequently, much as I would like to have a unanimous opinion, I cannot agree to one which contains the [statement] quoted above."

The Justice also objected to language in the Burger draft which referred to the location of schools so there would be black schools for black neighborhoods and white schools for white neighborhoods. "This sounds," Black wrote, "as though there can be something unconstitutional about sending pupils to a school in their neighborhood, closest to their homes. This could be laid aside as has been done on other subjects at the request of others of the Brethren."

Justice Black then criticized the "burden" paragraph, which "state[s] that where schools are substantially disproportionate in their racial composition and that all or predominately [*sic*] all are of one race, they have the burden of showing that such school assignments are genuinely nondiscriminatory." The Brennan March 23 letter had also referred to this paragraph. But whereas Brennan had wanted to strengthen the burden requirement, Black wanted it eliminated. As the latter saw it, the statements in the Burger draft "seek to change in a court opinion the rules of equity fixing the burden of proof in equity cases. This part of the opinion, like others that purport to set out constitutional 'guide lines,' in my judgment constitutes lawmaking which should only be done by the Congress. I would eliminate those statements from the opinion."

Black's last demand was the elimination from the opinion of Part V(3), on remedial altering of attendance zones, which had indicated that "affirmative action in the form of remedial altering of attendance zones" might be ordered, even if "the pairing and grouping of non-contiguous zones" was required. Black declared that this "seems to authorize the Court to order pupils to be transported from opposite ends of the city."

It was one thing for the school authorities to institute an extensive busing system. But, the Justice asserted, it was beyond judicial power to do so:

> I do gravely doubt this Court's constitutional power, however, to compel a State and its taxpayers to buy millions of dollars worth of busses to haul students miles away from their neighborhood schools and their homes. Such an order by this Court appears not only to be "bizarre," as this Court suggests, but actually unconstitutionally requiring the States to spend money. In my judgment such an order would amount to the federal courts actually taking over the operation of the State's public schools but at the expense of the States.

Black concluded his letter by stating that if the passages to which he objected were eliminated, "as the Court has done for others as to other parts of the opinion, I shall be ready to agree. If you cannot conscientiously agree to eliminate these parts, however, I shall most likely file a separate opinion, a draft of which I send you with this note." Even in a case such as this, "I am of the opinion that it would be a mistake to give the appearance of a unanimity on the Court which does not actually exist."

Black Dissent

The Black letter was accompanied by a sharp draft dissent of four typed pages, headed "Mr. Justice Black, concurring and dissenting." The draft stated that, though the Justice "would like to be able to join the Court's opinion, I regret that there are statements and possible implications in it with which I cannot agree." In particular, the dissent disagreed with the Court's handling of the racial balance issue. "Unlike the Court, I do not believe it constitutionally permissible for federal courts to use racial balance as a 'guideline' for remedying past discrimination and I do not believe that there is a constitutional 'presumption' against one-race schools. The Constitution sets its own guideline—*no discrimination on account of race.*"

Justice Black objected strongly to the use of busing to achieve racial balance:

I strongly disagree with any implication that may be found in the Court's opinion that federal courts may order school boards to transport students across cities to balance schools racially or to eliminate one-race schools which have resulted from private residential patterns rather than school-board imposed segregation. I particularly disagree with the Court's affirmance of an order that may require the taxpayers of North Carolina to spend millions of dollars for school busses to transport students miles across their home city apparently for the purpose of achieving substantial racial balance. The Constitution, in my judgment, does not require racial balance.

If Justice Black had issued this dissent, the effect would have been unfortunate. The Alabaman was the only Southerner on the Court and the quondam leader of the Court's liberal wing. Until then, he had been as firm as anyone in his view (as he put it in another portion of his *Swann* draft dissent) "that state-imposed racial segregation in public schools cannot, under the Constitution, be tolerated" and that "school boards throughout the country [must] comply with the mandate of *Brown* v. *Board of Education* . . . and fulfill their responsibilities to all Americans *now*." For Black to breach the unity that had prevailed in school desegregation cases since *Brown* would send precisely the wrong signal to both the South and the rest of the country.

It is, however, most unlikely that Black ever intended his draft dissent to be more than a bargaining ploy. Not unnaturally, the Justice wanted to secure what changes he could in the Burger draft passages which he disapproved. To prevent his dissent, he knew, the Chief Justice would do his best to meet at least some of the Black objections. Even more important, the threat of a Black dissent would prevent further rewriting of the *Swann* opinion to make it more palatable to Judge McMillan's strong supporters. In his discussions with the others, the Chief Justice could use Black's intransigence as the reason for his refusal to make the changes still sought by the five who favored Justice Brennan's suggestions. Even Brennan could conclude that the Burger draft joined by all was still preferable, with all its imperfections, to a further refining process that would produce the sharp Black dissent.

Sixth Draft

The threatened Black dissent was the catalyst that led to a speedy termination of the *Swann* decision process. The Justices had already come close to agreeing on the fifth draft opinion by the Chief Justice. But absent the threatened Black dissent, Justice Brennan might still have held out for further changes to strengthen the opinion. However, now he was willing to join the others even though this final Burger draft would contain only cosmetic modifications. Justice Black, on his side, would surely secure some concessions from the Chief Justice so as to stop his threatened dissent. More important, along with the Chief Justice, he had prevented the far stronger opinion in support of Judge McMillan that the majority had desired. When he compared the straightforward Stewart draft that the majority would have voted for with the still-fuzzy Burger opinion, he could conclude that his view had won as much as it had lost. After all, the Chief Justice's draft had prevented a stronger endorsement of racial balance and unlimited busing to achieve it. The Burger draft's retention, over the objection of Justice Brennan and some of the others, of the statement "that the limits on time of travel will vary with many factors, but probably with none more than the age of the students" might lead the lower courts to think twice before ordering massive busing of very young children.

On April 8 the Chief Justice circulated the sixth and final draft of his *Swann* opinion. It was accompanied by a *Memorandum to the Conference*, which stated, "As our respective files will show, I believe I have demonstrated a flexible attitude, even down to using words of others when I saw no real difference and preferred my own. The new draft in-includes more changes made at the suggestion of various members of the Court."

The principal points to be noted in the new draft were listed in the Burger memo. First, "I have accepted in a somewhat modified form Justice Brennan's and Justice Douglas' reservation section." The new passage on the matter read, "We do not reach in this case the question whether a showing that school segregation is a consequence of other types of state action, without any discriminatory action by the school authorities, is a constitutional violation requiring remedial action by a school desegregation decree. This case does not present that question and we therefore do not decide it."

This was not exactly the language suggested by the Brennan March 23 letter, but it was close enough to enable Justices Brennan and Douglas to conclude that their objection had been met. The sixth draft did not, however, follow a broader Brennan suggestion. In the margin of his copy of the new Burger draft, next to the paragraph containing the passage just quoted, Justice Brennan wrote, "Rejects our suggestion to omit this & ¶ on top of next page."[4] Brennan had suggested deleting both the paragraph referring to forms of discrimination other than school segregation and the next one stating that the Court's objective "does not and cannot embrace all the problems of racial prejudice." As the Brennan marginal comment shows, the Chief Justice refused to accept the suggestion.

The Burger memorandum also referred to the burden placed upon school boards by the existence of one-race schools: "As for the 'burden' discussion . . . I continue to prefer my own language, rather than Justice Brennan's." The language which the Chief Justice preferred had, as has been seen, been drafted by Justices Stewart and Harlan, so that it was really their phrasing which he preferred rather than that suggested by Justice Brennan. The retention of the "burden" discussion from the fifth draft was also rejection of the Black suggestion on the matter. As the Burger memo explained, "Justice Black would eliminate the 'burden' altogether, but that bridge was probably crossed in what was said in *Green* (391 U.S. at 439) on this matter."

The Burger memo then stated that the new draft had "adopted Justice Brennan's compromise for the first paragraph of Part IV." This was a misprint, since no changes were made in the paragraph to which the memo referred. The Chief Justice meant to refer to Part VI, where he had followed the Brennan suggestion to reinsert the sentence "On the facts of this case, we are unable to conclude that the order of the District Court is not reasonable, feasible and workable." Nevertheless, the draft continued to reject Justice Brennan's more important recommendation (originally made by Justice Harlan) that Part VI should not begin by referring to the court of appeals "reasonableness" test and the *Green* test.

On the ground that it "is not agreeable to others," the Chief Justice also rejected Justice Marshall's ssuggestion that the paragraph on majority-to-minority transfer provisions should be removed. The new draft did, however, make one change in the paragraph. The prior drafts had started the paragraph with the statement "An optional majority-to-minority transfer provision has long been recognized as an obvious and necessary part of every desegregation plan." The sixth draft substituted "a useful" for the words "an obvious and necessary." In his covering memo, the Chief Justice wrote, "I hope my suggested compromise will be acceptable, since we cannot affirm the order in *Swann* without saying something on the transfer plan adopted by Judge McMillan."

The Chief Justice next referred to some of Justice Black's other suggestions, saying, "As Justice Black's memorandum correctly suggests, compromise is a two-way process." The new draft met the Black objection to the statement that the Civil Rights Act was enacted "to *expand and* define the role of the Federal Government in the implementation of the *Brown I* decision" by deleting the first two italicized words. Also eliminated was the sentence "It is clear that Title IV of the Civil Rights Act of 1964 was not intended to limit the permissible remedies for a violation of the Fourteenth Amendment in school desegregation cases."

Another Black objection, to the paragraph on school construction to foster one-race schools, was met by omitting the sentence "In this case, the District Court made a finding, accepted by the Court of Appeals, that among the most important means used by the local authorities in the creation and maintenance of a dual system was 'locating and controlling the capacity of schools so that there would usually be black schools handy to black neighborhoods and white schools for white neighborhoods.' " The draft also made another change aimed at Justice Black. The passage on busing was modified by elimination of the sentence "The search for solutions is not aided by simplistic slogans for or against 'bussing,' as though the term described a uniformly invidious course of action."

Aside from some minor stylistic adjustments, the changes summarized in the Burger memo, which have been discussed, were the only alterations made in the sixth draft. In particular, the draft did not eliminate the section on remedial altering of attendance zones, as Justice Black had so strongly urged—though (presumably to mollify the Justice) a sentence was added at the end of the section: "Conditions in different localities will vary so widely that no rigid rules can be laid down to govern all situations."

In addition, the Burger memo referred to the draft's refusal to deal more fully with the resegregation issue, as some had suggested. "I am still of the view," the Chief Justice wrote, "that if we are not to suggest by way of dicta any direction on the so-called 'de facto' aspect we should avoid discussion of that issue now. We will not lack for cases on this in the future and it is apparent that we will not achieve unanimity on it now."

Final Approvals

At the end of the memorandum accompanying his sixth draft, the Chief Justice declared, "I do not prefer all of these changes," but had made them in order to secure a unanimous opinion. "I am sure it is unnecessary to say that if this objective cannot be achieved there will be no point in my using the language of others in preference to my own and I will naturally restore my own choice of language in the draft which will command a Court."

The Burger threat to restore his original language and secure a Court to support it was inconsistent with reality. As already stressed, at least five Justices had expressly indicated that they were prepared to follow the Stewart draft in preference to the first Burger drafts and there is no doubt that they would have strongly resisted attempts by the Chief Justice to revive his earlier versions of the *Swann* opinion. At the same time, the changes made in the opinion (particularly after the third draft had come out with a flat affirmance of Judge McMillan) had removed all their important objections to the original Burger draft. Even before the sixth draft, most of the Justices who supported McMillan had given up the struggle to secure further changes in the *Swann* opinion.

It is true that Justice Brennan had still tried to procure alterations in the Burger fifth draft, but even he now realized that he had obtained as much as he was going to. The sixth draft did, after all, meet some of the points that had been made in the Brennan March 23 "Dear Chief" letter. More important, Justice Brennan now had to weigh his remaining objections in the light of the threatened Black dissent. Whatever further refinements in the opinion he might want, they were certainly not worth securing if the price was to be publication of the draft dissent that Justice Black had circulated.

On April 9, the day after the Chief Justice circulated his sixth draft, Justice Brennan sent a "Dear Chief" letter in which he declared, "I am happy to join your 6th Draft. . . . I appreciate your consideration of my suggestions." Justices Marshall, Stewart, Harlan, White, Douglas, and Blackmun also agreed to the opinion, and so, on April 15, did Justice Black. As already indicated, he, too, had decided that it was safer to join the opinion than to carry out his threat to dissent if all his objections were not met. The latter would not only mean a breach in unanimity in a desegregation case by one of the leaders in the post-*Brown* judicial struggle; it might lead to an even stronger Court opinion in support of Judge McMillan's far-reaching order—the last thing, of course, that Justice Black wanted.

On April 14 the Chief Justice sent around thirteen pages of the opinion on which, he wrote in his covering *Memorandum to the Conference,* "minor 'touch up' changes have been made." As explained in the memo, "Largely they are paragraph adjustment to facilitate reading, the elimination of a few superfluous words, the insertion of others for smoother reading." Two small substantive changes were also made on the relevant pages: deletion of a sentence at the end of Part I (originally suggested by Justice Brennan) that read, "This case involves a metropolitan area"; and insertion of the italicized word in the following sentence at the end of the section on racial balance: "In sum, the *very* limited use made of mathematical ratios was within the equitable remedial discretion of the District Court."

The Chief Justice concluded his April 14 memo: "I believe we should have no difficulty in the final printing and the release date now turns on whether other opinions will be written."

No other opinions were, however, to be issued in the case. The day after the Burger memo, as already seen, Justice Black sent his note to the Chief Justice expressly joining the *Swann* opinion. Justices Douglas and Stewart, who had also prepared draft dissents, had decided as well that unanimity was more important than an opinion that was in complete accord with their views. The last Burger draft could, therefore, come down as the unanimous *Swann* opinion, without even a single concurring or dissenting voice to detract from its position as *the* opinion of the Court.

12

Swann Song

Tuesday morning, April 20, 1971. The Supreme Court chamber was packed, as usual. There was an air of expectancy, for the Court scuttle-butt had it that the *Swann* school-busing decision was about to come down. When the large clock behind the bench stood at precisely 10 A.M., the red curtains beneath it parted and, led by Chief Justice Burger, the Justices took their places. After the time-honored intonation, they sat down in their plush black-leather chairs, looking (as Justice Van De-vanter once said) "like nine black beetles in the temple of Karnak." The Chief Justice leaned forward and began in his mellow bass: "I have for announcement the judgment of the Court in No. 281—*Swann v. Char-lotte-Mecklenburg Board of Education.*"

During its early years' the Court's schedule was relatively informal. Conferences were held whenever the Justices decided they were neces-sary—sometimes in the evenings and on weekends and often in the boardinghouse where the Justices lodged. Similarly, decisions were an-nounced whenever they were ready. In 1857, the Court began the tradi-tion of announcing decisions on Monday; hence the press characteriza-tion of "decision Mondays." The practice continued for over a century. Then, under Chief Justice Warren's prodding, the Court announced in 1965 that "it will no longer adhere to the practice of reporting its deci-sions only at Monday sessions and that in the future they will be re-ported as they become ready for decision at any session of the Court."[1] Not everyone considered the change a sign of progress. The *New York Times* had reported that newspapermen had always admired the Court precisely because of its refusal to arrange its work for the convenience of the press.[2]

Since "decision Mondays" had thus come to an end a few years

185

earlier, Chief Justice Burger could announce the *Swann* decision on the Tuesday morning. The Supreme Court practice has always been to have decisions announced orally in open court. The Justice who wrote the opinion announces the Court's decision, and Justices who wrote concurring or dissenting opinions may state their views as well. It is up to the individual opinion writer whether to read the opinion, or summarize it, or simply to announce the result and state that a written opinion has been filed.

Decision day in *Brown* v. *Board of Education*[3] had been one of the most dramatic events in Supreme Court history. Chief Justice Warren had read the unanimous opinion striking down school segregation to a courtroom pervaded by tension.[4] The Warren tone may have been colorless, but his voice sounded a clarion for the civil rights movement that was to transform the American society.

Now, in *Swann,* Chief Justice Burger made what *Newsweek* termed "by far the most momentous Court pronouncement on school segregation since the landmark Brown decision of 1954."[5] Burger has not, however, followed the practice of his predecessor in reading opinions on decision day. In recent years, indeed, the standard Burger procedure has been simply to announce the judgment with no elaboration or summary. At the time of *Swann,* he had not departed so fully from the prior practice. The Chief Justice announced *Swann* by briefly summarizing the opinion and stating that the judgment of the court of appeals was affirmed so far as it affirmed the judgment of the district court and that the order of the district court was also affirmed.

The opinion that came down in the Chief Justice's name was, of course, anything but the *Swann* opinion that Burger had sought to deliver. "There is a lot of conflicting language here," Judge Griffin B. Bell, of the U.S. Court of Appeals for the Fifth Circuit, told a *Newsweek* reporter after he had read the *Swann* opinion. "It's almost as if there were two sets of views laid side by side."[6]

Judge Bell spoke more accurately than he knew. The final *Swann* opinion was a composite, containing both the lukewarm views of the Chief Justice in whose name it was delivered and the strong views in support of desegregation advocated by the Justices who favored a clear affirmance of Judge McMillan. The result, as the *New Republic* put it, was "a negotiated document looking in more than a single direction."[7]

The unsatisfactory nature of the Burger opinion did not, however, obscure the fact that the *Swann* decision was a categorical affirmance of Judge McMillan's far-reaching desegregation order. It was most unusual for the Supreme Court to affirm a court of appeals judgment only "as to those parts in which it affirmed the judgment of the District Court" and then, to emphasize the kudo conferred on the district judge, to state expressly that his order "is also affirmed."⁸ This language (which the Chief Justice had resisted until his third draft) was a clear signal that the Justices were solidly behind McMillan's tough integration stance.

McMillan himself—in many ways the central figure in the drawn-out and emotionally packed case—was at work in his chambers in Charlotte when Chief Justice Burger announced the *Swann* decision. Shortly after 10:30 A.M. the telephone rang. David E. Gillespie, the associate editor of the *Charlotte Observer,* was on the line and asked to speak to the judge. His secretary said that he was in conference. Gillespie asked if the judge knew that the Supreme Court had ruled on the case, upholding his order. The secretary then said that she would tell the judge to pick up the phone and get the news. She later said, "I figured that even if the roof fell in on me, I'd interrupt him."

Gillespie told McMillan that the wire services had just reported the news of the Supreme Court ruling. "I felt," Gillespie reported, "that he was momentarily overwhelmed by the news." When he had recovered his composure, the judge declared, "If the court did what's been reported, they've done the classic job of an appellate court under the rule of law."⁹

Crossing the Bridge

"He crossed the bridge with this," a Justice who had participated in the *Swann* decision told me. "He tried to take over the Court. He didn't succeed and never tried this blatantly again."

The Justice was referring to Chief Justice Burger's efforts to mold the *Swann* decision and opinion to reflect his own more restricted view of desegregation power even though that view was opposed by a majority of the Justices. In this sense, the *Swann* decision process involved a power struggle between the Chief Justice and those who favored a strong affirmance of Judge McMillan. As the Justice quoted above put it to me, "The Chief Justice was determined to carry the day, and we were equally determined he wouldn't—and we had the votes."

His lack of votes induced Burger to agree to rewrite and ultimately to issue a *Swann* opinion that the others were willing to accept. But the final opinion still was not as forthright as it would have been had it been written by one of McMillan's firm supporters. Describing the redrafting process, the same Justice said, "We tried to save as much as we could [from the Burger drafts] so that the Chief Justice wouldn't lose face—but if he had to lose face, so be it." As it turned out, it never came to the latter alternative. Enough of the Burger drafts was saved to enable the opinion to be issued under the Chief Justice's name, though this made the imprimatur placed on strong integration orders weaker than it would have been had the Stewart draft been substituted for the Burger version.

At the same time, the majority Justices could feel that they had accomplished a great deal in ensuring that the original Burger view did not prevail. The extent of their accomplishment can be seen from a comparison of the first Burger *Swann* draft, circulated on December 8, 1970 (Appendix A), with the final opinion that was issued four and a half months later (Appendix B).

Had the first Burger draft come down as the final *Swann* opinion, it would have marked a serious setback for enforcement of civil rights. The negative tone taken by the Chief Justice to Judge McMillan's busing order would have sent a clear message that the Warren Court's expansive approach to desegregation would no longer be followed. In its place there was to be a restrictive attitude toward the remedial powers of the federal courts that would render them powerless to achieve more than cosmetic changes in segregated school systems.

Instead, the signal sent by *Swann* was one of solid support for judges like McMillan who took vigorous measures to end dual school systems. The discretion which the Chief Justice had originally sought to deny them was now confirmed in broad language by the Supreme Court. They were told that they had broadside authority to order any remedies, including busing, to root out "all vestiges of state imposed segregation."[10] With *Swann*, the federal courts were, more firmly than ever, engaged in the business of actively ensuring the achievement of desegregation.

"We Sold the Chief Justice"

Julius Chambers, the attorney for the *Swann* plaintiffs, was told about the Supreme Court decision by his minister. He did not say a word. Later he declared, "That's great," and admitted he was "elated." He stressed to one of his associates that the "decision was unanimous. And the opinion was written by Burger. We sold the Chief Justice."[11]

Of course, Chief Justice Burger had not been "sold" by the Chambers argument. He had resisted, as long as he could, the affirmance of Judge McMillan and the seal of approval placed upon his extensive busing order. Nor did the fact that he yielded to the opposition within the Court and finally issued a *Swann* opinion so different from his original draft mean that he had changed his mind on the merits. He went out of his way to demonstrate this publicly a few months after the *Swann* decision in a case involving the schools in Winston-Salem, North Carolina.

The Winston-Salem Board of Education had applied to Burger, in his capacity as supervising circuit justice for the fourth circuit, for a stay of a desegregation plan adopted by the district court. The district judge had relied on the *Swann* decision to order extensive busing throughout the school district. The Chief Justice issued an opinion on August 31, 1971, denying a stay on the technical ground that the record was not clear on whether the *Swann* holding had been correctly applied.

The Burger opinion did, however, state the Chief Justice's interpretation of *Swann,* particularly what he termed its "explicit language as to a requirement of fixed mathematical ratios or racial quotas and the limits suggested as to transportation of students." Burger quoted the *Swann* opinion, adding the emphasis to make his restrictive approach clearer:

> If we were to read the holding of the District Court to require, as a matter of substantive constitutional right, any particular degree of racial balance or mixing, *that approach would be disapproved and we would be obliged to reverse.* The constitutional command to desegregate schools does not mean that every school in every community must always reflect the racial composition of the school system as a whole.[12]

"Nothing could be plainer," the Chief Justice wrote, "or so I had thought, than *Swann's* disapproval of the 71%–29% racial composition found in the *Swann* case as the controlling factor in assignment of pupils, simply because that was the racial composition of the whole school system." This was true even though he conceded that "we had noted the necessity for a district court to determine what in fact was the racial balance as an obvious and necessary starting point to decide whether in fact any violation existed."[13]

But what Burger failed to note was that the *Swann* opinion had approved "the use made of mathematical ratios" as "a starting point *in the process of shaping a remedy.*"[14] By omitting the reference to "shaping a remedy," the Chief Justice had virtually reversed the *Swann* approval of the use made by Judge McMillan of racial balance as a starting point in the working out of his desegregation plan. In effect, Burger was asserting the view that he had taken in his first *Swann* draft as that adopted by the Court in the final *Swann* opinion.

The *Winston-Salem* opinion also stressed what the Chief Justice called the "limits" imposed by *Swann* on busing, quoting the *Swann* language on "the time or distance of travel [being] so great as to either risk the health of the children or significantly impinge on the educational process," and emphasizing "the age of the students" as a determinative factor.[15] Here Burger was going back to the approach followed by the court of appeals in the Swann case—the approach that had been repudiated by Swann's reversal of the limitations imposed by the court of appeals on Judge McMillan's busing order.

The other Justices resented the Burger effort to redefine the Swann decision to make it accord with the view taken in his first Swann draft; but there was nothing they could do about it in the Winston-Salem case itself. By denying the stay, the Chief Justice had allowed the district court's order to go into effect. By voting to hear the case on the merits, the Justices would only cast doubt on that result. Not surprisingly, then, the Court denied certiorari, allowing the Winston-Salem busing order to stand. Also left untouched was the Chief Justice's Winston-Salem opinion.

But his restrictive interpretation in that opinion was no more able to control the law on the scope of desegregation orders than were his first drafts in the Swann case. As the Justices who strongly supported Judge McMillan had realized, the crucial consideration was the categorical affirmance of the Swann busing order. Nothing that was said in the Burger Winston-Salem opinion could blur the effect of the signal Swann had given to the federal judges, a signal that led to widespread busing orders in courts across the South in response to the Swann decision.[16]

Closing of the File

After Judge McMillan issued his school-busing order, Time reported that he became "a pariah to many in the community. Though he is an avid golfer (with a 9 or 10 handicap), rumor had it that he was unable to pick up a foursome."[17]

When he spoke to me, McMillan referred to the Time account that "I couldn't get up a golf game," declaring, "I love golf." After the story, he recalled, "I counted up the people who played with me. When I got to sixty, I stopped."

Despite his denial, there was something poignant about the manner in which the judge went out of his way to bring up the Time article. So many years later, the judge was still affected by the ostracism he suffered at the time of the Swann case. Because of his busing order he was picketed, threatened, and hanged in effigy.[18] Julius Chambers, as we saw in Chapter 1, suffered even more severely. His home, office, and car were bombed[19]—and all because he was the principal attorney for the Swann plaintiffs.

In 1981, both McMillan and Chambers were the guests of honor at a testimonial dinner in Charlotte. "Everybody who's anybody is here to-night," said a local newspaperman of the integrated audience of 300 that turned out to honor the white judge who had ordered the school busing and the black lawyer who had argued for it.[20] The *Swann* busing had been bitterly resisted both by parents' groups and civic and school officials. Now the current school board had canceled its own meeting to attend the McMillan-Chambers dinner.

The dinner led to an editorial in the *Charlotte Observer* on the decade of busing that had resulted from Judge McMillan's order. Because of busing, according to the editorial, Charlotte was no longer racially polarized. "Schools are no longer black or white, but are simply schools. As a result, the racial composition of surrounding areas is not as critical as it once was. The center city and its environs are a healthy mixture of black and white neighborhoods."[21]

One who visits Charlotte is bound to be impressed by the integration accomplished in its school system. What is more, the quality of education in Charlotte's integrated schools is at least as high as that in similar educational districts throughout the country. Most observers believe that the high standards maintained in Charlotte's schools is a direct result of Judge McMillan's desegregation orders and their ultimate acceptance by the bulk of the community.

Community acceptance, however, did not come easily. As seen in Chapter 1, there was bitter opposition to the integration ordered by Judge McMillan. This was vividly recalled in a statement by one of the first blacks to attend a formerly all-white Charlotte school: "that morning when Daddy let me out of the car and . . . I started walking toward the school, a lot of people were just pushing and shoving and calling me names and throwing things, spitting on me and saying . . . 'Nigger, go back to Africa.' "[22]

The opposition to school busing reached its Charlotte crescendo with the Supreme Court's *Swann* decision. According to a reporter, "The desegregation of the Charlotte-Mecklenburg schools through court-ordered busing temporarily wrecked this community."[23] With the Supreme Court decision, however, Charlotte came to realize that McMillan's orders were not going to be undone. The community trauma began to be replaced by a desire to work out an orderly desegregation process acceptable to all but extremists. The moderate elements in the community joined with the education authorities to draft a new pupil assignment plan.

The work of the Citizens Advisory Group (the organization that worked with the school board for the purpose) was recently summarized by Judge McMillan: "I decided that perhaps this group, some of whose recommendations had been printed in the paper, would be well to act as a go-between between the court and the board. . . . So I asked

[them] to work up a plan with the assistance of the school staff and present it to the court." Then, McMillan went on, "they worked up a plan which is the core of the pupil assignment plan today. That made it possible for the board to withdraw gracefully from the war."[24]

The Supreme Court decision had also convinced political and business leaders in Charlotte that further resistance to the McMillan orders was futile. In 1973, the chairman of the Board of County Commissioners asked Judge McMillan what it would take to get him out of the school business. The judge answered that all it would take was a school board willing to obey the law of the land.[25] The political powers-that-be in the community then saw to it that a new school board majority was elected whose members were willing to accept effective desegregation, including school busing. The new board agreed to the plan worked out by the Citizens Advisory Group and the school staff. That, in turn, says McMillan, "made it possible for me to approve the plan, which was good in the principles and in fact. . . . I'd say that that . . . more than any other single factor brought the controversy to a close."[26]

In 1974, Judge McMillan entered an order approving the revised plan and expressing appreciation to the school board, the Citizens Advisory Group, and the school staff people and others who had worked on the plan.[27] A year later, the judge issued what he termed "FINAL ORDER (SWANN SONG)." In it he ended the *Swann* litigation. After noting the "more positive attitude" of the new board and the fact that they were "actively and intelligently addressing" the desegregation problem, McMillan stated that he was acting "to leave the constitutional operation of the schools to the Board, which assumed that burden after the latest election."

"With grateful appreciation," the last McMillan *Swann* opinion concluded, "to all who have made possible this court's graduation from *Swann,* it is therefore
Ordered:

1. That this cause be removed from the active docket.
2. That the file be closed."[28]

Nine years later, in 1984, the National Education Association cited Charlotte as an example of a city that is making desegregation work.[29] The current situation in the Charlotte-Mecklenburg School District was described by the *New York Times* in an article on the twentieth anniversary of the *Brown* decision:[30] "It has been a dozen years since anyone has been elected to the Board of Education on an antibusing platform, and two of the nine board members are black. For the past four years, the county's students have scored above average in a national achievement test, and the gap in test scores between black and white students has been narrowing."[31] In 1983, Charlotte elected its first black mayor.[32]

And what of the Swanns, for whom the landmark Supreme Court decision was named? Reverend Darius Swann may have given his son's name as the lead plaintiff, but he and his family had left Charlotte a year before the case was reopened in 1968, and never returned. When the Supreme Court announced its decision, the family was in India, where the father was teaching and doing missionary work. They probably did not even hear of the decision and were completely unaffected by it.[33]

Notes

Abbreviations Used

BRW   Byron R. White
FF   Felix Frankfurter
FFH   Felix Frankfurter Papers, Harvard Law School
FFLC   Felix Frankfurter Papers, Library of Congress
HLB   Hugo L. Black
HLBLC   Hugo L. Black papers, Library of Congress
JMH   John M. Harlan
JMHP   John M. Harlan Papers, Mudd Manuscript Library,
         Princeton University
PS   Potter Stewart
TM   Thurgood Marshall
WEB   Warren E. Burger
WJB   William J. Brennan
WOD   William O. Douglas

Chapter 1

1. Letter to author, May 4, 1984.
2. 402 U.S. 1 (1971).
3. Court-ordered School Busing. Hearings before the Subcommittee on Separation of Powers of the Committee on the Judiciary, United States Senate, 97th Cong., 1st Sess., p. 511 (1982); hereafter cited as Hearings.
4. Charlotte Observer, July 12, 1975, p. 18A.
5. Hearings 512, 527.
6. Id. at 527, 528.
7. Id. at 528.
8. Id. at 511.
9. Id. at 528.
10. Id. at 528–529.

11. Southern Pacific Co. v. Jensen, 244 U.S. 205, 211 (1917).
12. See Peltason, Fifty-Eight Lonely Men: Southern Federal Judges and School De-
    segregation 4 (1961).
13. 347 U.S. 483 (1954); 349 U.S. 294 (1955).
14. 347 U.S. 483 (1954).
15. 349 U.S. 294 (1955).
16. Id. at 301. Italics added.
17. See Peltason, op. cit. supra note 12, at 8.
18. Peltason, op. cit. supra note 12.
19. Kluger, Simple Justice: The History of Brown v. Board of Education and Black
    America's Struggle for Equality 100 (1976).
20. All quoted statements for which a source is not given were made in interviews
    with the author.
21. Hughes, Fight for Freedom: The Story of the NAACP 123 (1962).
22. Id. at 146.
23. 371 U.S. 415 (1963). For a fuller account see Schwartz, Super Chief: Earl
    Warren and His Supreme Court—A Judicial Biography 450-452 (1983).
24. 372 U.S. 539 (1963). For a fuller treatment, see Schwartz, op. cit. supra note
    23, at 452-453.
25. Three Cities that Are Making Desegregation Work, Report of a National Educa-
    tion Association Special Study 24 (1984).
26. See Metcalf, From Little Rock to Boston: The History of School Segregation 98
    (1983).
27. National Law Journal, July 30, 1984, p. 24.
28. Ibid.
29. Op. cit. supra note 25, at 26.
30. Swann v. Charlotte-Mecklenburg Board of Education, 300 F. Supp. 1358, 1362
    (W.D.N.C. 1969).
31. Loc. cit. supra note 29.
32. Swann v. Charlotte-Mecklenburg Board of Education, 243 F. Supp. 667, 670
    (W.D.N.C. 1965).
33. Swann v. Charlotte-Mecklenburg Board of Education, 369 F.2d 29, 32 (4th Cir.
    1966).
34. 391 U.S. 430 (1968).
35. Id. at 437-438, 439.
36. 396 U.S. 19 (1969).
37. Id. at 20.
38. Swann v. Charlotte-Mecklenburg Board of Education, 300 F. Supp. 1358, 1363
    (W.D.N.C. 1969).
39. Brown v. Board of Education, 347 U.S. 483 (1954).
40. Hearings 528, 512, 511.
41. Id. at 529.
42. 402 U.S. at 6-7.
43. Swann v. Charlotte-Mecklenburg Board of Education, 300 F. Supp. 1358, 1367-
    1368 (W.D.N.C. 1969).
44. Hearings 531, 532.
45. Swann v. Charlotte-Mecklenburg Board of Education, 300 F. Supp. 1358, 1368
    (W.D.N.C. 1969).
46. William J. Waggoner, interview.
47. Hearings 512.
48. Swann v. Charlotte-Mecklenburg Board of Education, 300 F. Supp. 1358, 1372
    (W.D.N.C. 1969).
49. Ibid.
50. Hearings 512.

51. Swann v. Charlotte-Mecklenburg Board of Education, 300 F. Supp. 1358, 1360, 1373, 1386 (W.D.N.C. 1969).
52. Interview with author.
53. Hearings 529.
54. Alexander v. Holmes County Board of Education, 396 U.S. 19 (1969).
55. Swann v. Charlotte-Mecklenburg Board of Education, 300 F. Supp. 1358, 1382 (W.D.N.C. 1969).
56. Id. at 1381, 1386.
57. Swann v. Charlotte-Mecklenburg Board of Education, 306 F. Supp. 1291, 1299, 1298 (W.D.N.C. 1969).
58. Swann v. Charlotte-Mecklenburg Board of Education, 306 F. Supp. 1299 (W.D.N.C. 1969).
59. 396 U.S. 19 (1969).
60. Swann v. Charlotte-Mecklenburg Board of Education, 306 F. Supp. 1299, 1301 (W.D.N.C. 1969).
61. Id. at 1309.
62. Id. at 1313.
63. Swann v. Charlotte-Mecklenburg Board of Education, 300 F. Supp. 1358, 1386 (W.D.N.C. 1969).
64. Swann v. Charlotte-Mecklenburg Board of Education, 311 F. Supp. 265, 266 (W.D.N.C. 1970).
65. Swann v. Charlotte-Mecklenburg Board of Education, 431 F.2d 138, 148 (4th Cir. 1970).
66. Infra p. 168.
67. Swann v. Charlotte-Mecklenburg Board of Education, 318 F. Supp. 786, 799 (W.D.N.C. 1970).
68. Swann v. Charlotte-Mecklenburg Board of Education, 311 F. Supp. 265, 270 (W.D.N.C. 1970).
69. Ibid.
70. 402 U.S. at 8.
71. Id. at 9.
72. Id. at 10.
73. Swann v. Charlotte-Mecklenburg Board of Education, 311 F. Supp. 265, 268 (W.D.N.C. 1970).
74. Hearings 522.
75. This was essentially the school board's plan for secondary schools, with modifications.
76. Swann v. Charlotte-Mecklenburg Board of Education, 311 F. Supp. 265, 268 (W.D.N.C. 1970).
77. Barrows, School Busing: Charlotte, N.C.; Atlantic Monthly 17, 18 (Nov. 1972).
78. Swann v. Charlotte-Mecklenburg Board of Education, 300 F. Supp. 1358, 1369 (W.D.N.C. 1969).
79. Barrows, supra note 77, at 18.
80. Swann v. Charlotte-Mecklenburg Board of Education, 306 F. Supp. 1299, 1312 (W.D.N.C. 1969).
81. Id. at 1313, as quoted in Swann v. Charlotte-Mecklenburg Board of Education, 402 U.S. at 23.
82. Id. at 24–25.
83. Swann v. Charlotte-Mecklenburg Board of Education, 318 F. Supp. 786, 793 (W.D.N.C. 1970).
84. See id. at 792.
85. See Swann v. Charlotte-Mecklenburg Board of Education, 311 F. Supp. 265, 277–278 (W.D.N.C. 1969).

86. Swann v. Charlotte-Mecklenburg Board of Education, 318 F. Supp. 786, 792 (W.D.N.C. 1970).
87. Barrows, supra note 77, at 19.
88. March 30, 1970, p. A3.
89. Barrows, supra note 77, at 19–20; Time, May 3, 1971, p. 40; Charlotte Observer, July 12, 1975, p. 19A.
90. New York Times, February 5, 1971, p. 6.
91. National Law Journal, July 30, 1984, p. 1.
92. Id. at 24.
93. Hearings 520, 522.
94. Swann v. Charlotte-Mecklenburg Board of Education, 318 F. Supp. 786, 792 (W.D.N.C. 1970).
95. Swann v. Charlotte-Mecklenburg Board of Education, 306 F. Supp. 1291, 1293 (W.D.N.C. 1969).
96. Swann v. Charlotte-Mecklenburg Board of Education, 398 U.S. 978 (1970).
97. Swann v. Charlotte-Mecklenburg Board of Education, 431 F.2d 135 (4th Cir. 1970).
98. Swann v. Charlotte-Mecklenburg Board of Education, 431 F.2d 138, 145, 140 (4th Cir. 1970).
99. All letters by fourth circuit judges are from the Simon E. Sobeloff Papers, Library of Congress.
100. Swann v. Charlotte-Mecklenburg Board of Education, 431 F.2d 138, 148, 156 (4th Cir. 1970).
101. Id. at 155, 156.
102. Simon E. Sobeloff–Harrison [Winter], May 20, 1970.
103. Ibid.
104. Albert V. Bryan–Judge Winter, May 21, 1970.
105. Ibid.
106. Swann v. Charlotte-Mecklenburg Board of Education, 431 F.2d 138, 160 (4th Cir. 1970).
107. Washington Post, March 30, 1970, p. A3.
108. Bartlett's Familiar Quotations 721 (15th ed. 1980).
109. New York Times, March 25, 1970, p. 1.

Chapter 2

1. Schwartz, Inside the Warren Court 4 (1983).
2. FF–Fred M. Vinson, n.d. FFLC.
3. 1 Bryce, The American Commonwealth 274 (1917).
4. FF–Stanley Reed, April 13, 1939. FFLC.
5. Congressional Quarterly's Guide to the United States Supreme Court 754 (1979).
6. See Schwartz, A Basic History of the U.S. Supreme Court 172 (1968).
7. FF–Harold H. Burton, January 31, 1956. FFLC.
8. FF–WJB, March 27, 1958. FFLC.
9. See Frankfurter, Of Law and Men 133 (1956).
10. 20 Wall. VIII, X (U.S. 1874).
11. Douglas, The Court Years 1939–1975, 222 (1980).
12. 396 U.S. 19 (1969).
13. 3 Beveridge, The Life of John Marshall 15–16 (1919).
14. 9 The Writings of James Madison 116 (Hunt ed. 1910).
15. 20 Wall. VIII, X (U.S. 1874).
16. See Haskins and Johnson, Foundations of Power: John Marshall 1801–15, 385–386, 384 (1981).
17. Fairman, Reconstruction and Reunion 1864–88, 66 (1971).

18. In a May 26, 1984, letter to the author, Professor Fairman indicates that he "spoke quite incautiously" and the practice under Chase may have been more "in doubt" than his quoted statement indicates.
19. Hughes, The Supreme Court of the United States 58–59 (1928).
20. See Washington Post, July 4, 1972, p. A1; Douglas, op. cit. supra note 11, at 232; Woodward and Armstrong, The Brethren: Inside the Supreme Court 66, 170, 179, 417 (1979).
21. 410 U.S. 113 (1973). See Washington Post, July 4, 1972, p. A1.
22. Quoted in Woodward and Armstrong, op. cit. supra note 20, at 187. See Washington Post, July 4, 1972, p. A10.
23. Lloyd Corp. v. Tanner, 407 U.S. 551 (1972).
24. Quoted in Woodward and Armstrong, op. cit. supra note 20, at 180.
25. Peters v. Hobby, 349 U.S. 331 (1955).
26. Gonzalez v. Freeman, 354 F.2d 570 (D.C. Cir. 1964).
27. JMH–WEB, May 21, 1970. JMHP.
28. 1970 WEB file. JMHP.
29. WEB–JMH, n.d. 1971 WEB file. JMHP.
30. JMHP.
31. New York Times, December 14, 1969, News of the Week, p. 9; Washington Post National Weekly Edition, January 2, 1984, p. 24.
32. WEB, Memorandum, June 1971. JMHP.
33. JMHP.
34. Compare, Douglas, op. cit. supra note 11, at 232.
35. JMHP.
36. New York Times, January 17, 1965, section VI, p. 58.
37. Quoted in The Supreme Court under Earl Warren 129 (Levy ed. 1972).
38. Dunne, Hugo Black and the Judicial Revolution 85 (1977).
39. Op. cit. supra note 37, at 135.
40. The Dictionary of Biographical Quotation 79 (Kenin and Winette eds. 1978).
41. Douglas, Go East Young Man: The Early Years 450 (1974).
42. Gerhart, America's Advocate: Robert H. Jackson 274 (1958).
43. Supra note 12.
44. Fortas, quoted in The Fourteenth Amendment Centennial Volume 34 (Schwartz ed. 1970).
45. Holmes, The Path of the Law, 10 Harv. L. Rev. 457, 469 (1897).
46. Dunne, op. cit. supra note 38, at 414.
47. 5 Burke, Works 67 (rev. ed. 1865).
48. Woodward and Armstrong, op. cit. supra note 20, at 97.
49. Griswold v. Connecticut, 381 U.S. 479 (1965).
50. Bell v. Maryland, 378 U.S. 226 (1964).
51. Mercedes Douglas–HLB, June 29. HLBLC.
52. Hugo Black, Jr., My Father: A Remembrance 239 (1975).
53. HLB–Alan Washburn, December 17, 1958. HLBLC.
54. Frankfurter, op. cit. supra note 9, at 133.
55. 163 U.S. 537 (1896).
56. WEB–JMH, n.d. JMHP.
57. JMH–WEB, June 9, 1970. JMHP.
58. FF–JMH, July 6, 1956. FFH.
59. JMH–FF, July 28 [1956]. FFH.
60. See Schwartz, Super Chief: Earl Warren and His Supreme Court–A Judicial Biography, Chapter 9 (1983).
61. Douglas, op. cit. supra note 11, at 250.
62. Clayton, The Making of Justice: The Supreme Court in Action 217 (1964).
63. Jacobellis v. Ohio, 378 U.S. 184, 197 (1964).

64. Clayton, op. cit. supra note 62, at 218.
65. Woodward and Armstrong, op. cit. supra note 20, at 47.
66. Douglas, op. cit. supra note 11, at 251.
67. Henry J. Friendly–FF, January 9, 1962. FFLC.
68. Henry J. Friendly–FF, February 2, 1962. FFLC.
69. Woodward and Armstrong, op. cit. supra note 20, at 48.
70. WEB file, 1971. JMHP.
71. FF–Sherman Minton, September 19, 1955. FFH.
72. Supra note 35.
73. Stevens, The Life Span of a Judge-Made Rule, 58 N.Y.U. L. Rev. 1(1983).
74. 369 U.S. 186 (1962).
75. Warren, The Memoirs of Earl Warren 306 (1977).
76. Alexander v. Holmes County Board of Education, 396 U.S. 19 (1969).
77. [WOD–HLB], June 17, 1963. HLBLC.

Chapter 3

1. 347 U.S. 483 (1954).
2. 163 U.S. 537 (1896).
3. Id. at 551.
4. See Schwartz, Super Chief: Earl Warren and His Supreme Court—A Judicial Biography 86 (1983).
5. 347 U.S. at 494–495.
6. Id. at 494, note 11.
7. See Schwartz, The Unpublished Opinions of the Warren Court ch. 11 (1985). The Warren draft Brown opinions are reprinted, id. at 451, 463.
8. This theme is developed in Cahn, Jurisprudence, 1954 Annual Survey of American Law 809.
9. Loc. cit. supra note 4.
10. Quoted, id. at 111.
11. 347 U.S. at 495.
12. Schwartz, op. cit. supra note 4, at 120.
13. Id. at 116.
14. Ibid.
15. Id. at 91.
16. Id. at 115.
17. Brown v. Board of Education, 349 U.S. 294, 299 (1955).
18. Id. at 300–301.
19. Id. at 301.
20. See Schwartz, op. cit. supra note 4, at 121–122.
21. Lusky, Racial Discrimination and the Federal Law: A Problem in Nullification, 63 Col. L. Rev. 1163, 1172, n.37 (1963).
22. Black, My Father: A Remembrance 209 (1975).
23. Compare Graglia, Disaster by Decree: The Supreme Court Decisions on Race and the Schools 35 (1976).
24. Schwartz, op. cit. supra note 4, at 115.
25. Id. at 93.
26. Ibid.
27. The phrase used by Justice Burton at the second Brown I conference. Ibid.
28. Id. at 118.
29. See Wilkinson, From Brown to Bakke—The Supreme Court and School Integration: 1945–1978, 81 (1979).
30. Briggs v. Elliott, 132 F. Supp. 776, 777 (E.D.S.C. 1955).
31. Compare Wilkinson, op. cit. supra note 29, at 82.

32. See Schwartz, A Basic History of the U.S. Supreme Court 153 (1968).
33. See Shuttlesworth v. Birmingham, 162 F. Supp. 372, 379 (N.D. Ala. 1958), affirmed, 358 U.S. 101 (1958).
34. See Note, 62 Col. L. Rev. 1448, 1477–1478 (1962).
35. Id. at 1453.
36. 238 F.2d 724 (4th Cir. 1956), cert. denied, 353 U.S. 910 (1957).
37. Id. at 728.
38. 353 U.S. 910 (1957).
39. 358 U.S. 101 (1958).
40. Shuttlesworth v. Birmingham, 162 F. Supp. 372, 384 (N.D. Ala. 1958).
41. Loc. cit. supra note 39.
42. Wilkinson, op. cit. supra note 29, at 85.
43. Shuttlesworth v. Birmingham, 162 F. Supp. 372, 382, n.11 (N.D. Ala. 1958).
44. Graglia, op. cit. supra note 23, at 40.
45. Compare Wilkinson, op. cit. supra note 29, at 95.
46. Green v. Roanoke School Board, 304 F.2d 118 (4th Cir. 1962).
47. Bickel, The Decade of School Desegregation: Progress and Prospects, 64 Col. L. Rev. 207 (1964), quoting id. at 124.
48. Supra p. 50.
49. McDonald, in From Brown to Bradley: School Desegregation 1954–1974, 29 (Browning ed. 1975).
50. 373 U.S. 668 (1963).
51. Supra note 36.
52. 373 U.S. at 676–677.
53. 373 U.S. 683 (1963).
54. Id. at 688–689.
55. Ibid.
56. Goss v. Board of Education, Mr. Justice Douglas, concurring, June ——, 1963, p. 2. JMHP.
57. Referring to Stell v. Savannah-Catham County Board of Education, 220 F. Supp. 667 (S.D. Ga. 1963).
58. WJB–Tom C. Clark, May 23, 1963. Tom C. Clark Papers, Tarlton Law Library, University of Texas.
59. 373 U.S. at 683.
60. Watson v. Memphis, 373 U.S. 526, 530 (1963).
61. Griffin v. County School Board, 377 U.S. 218, 221 (1964).
62. Allen v. County School Board, 266 F.2d 507 (4th Cir. 1959).
63. Quoted in Wilkinson, op. cit. supra note 29, at 98.
64. Griffin v. County School Board, 377 U.S. 218, 222 (1964).
65. Compare Wilkinson, op. cit. supra note 29, at 100.
66. 377 U.S. at 229.
67. Id. at 232.
68. Id. at 233.
69. Id. at 234.
70. Id. at 232.
71. Id. at 234.
72. 382 U.S. 103 (1965).
73. Compare Wilkinson, op. cit. supra note 29, at 96.
74. United States v. Jefferson County Board of Education, 372 F.2d 836, 853–854 (5th Cir. 1966).
75. Schwartz, op. cit. supra note 4, at 609.
76. 382 U.S. at 103.
77. Schwartz, op. cit. supra note 4, at 609–610.
78. Supra note 71.

79. 382 U.S. at 105.
80. 391 U.S. 430 (1968).
81. McKay, "With All Deliberate Speed" A Study of School Desegregation, 31 N.Y.U. L. Rev. 991, 1053 (1956).
82. Dunn, Title VI, The Guidelines and School Desegregation in the South, 53 Va. L. Rev. 42, 64–65 (1967).
83. Id. at 44.
84. Briggs v. Elliott, 132 F. Supp. 776, 777 (E.D.S.C. 1955).
85. Court-ordered School Busing. Hearings before the Subcommittee on Separation of Powers of the Committee on the Judiciary of the United States Senate, 97th Cong., 1st Sess., 543 (1982).
86. Dunn, supra note 82, at 65, n.119. See Green v. County School Board, 391 U.S. 430, 440, n.5 (1968).
87. 66 Landmark Briefs and Arguments of the Supreme Court: Constitutional Law 223 (Kurland and Kasper eds. 1975).
88. Schwartz, op. cit. supra note 4, at 704.
89. 358 U.S. 1 (1958). See Schwartz, op. cit. supra note 4, at 295–300. The Court's opinion was issued in the names of each of the nine Justices.
90. Supra note 72, issued as per curiam opinion.
91. Schwartz, op. cit. supra note 4, at 704. Compare Earl Warren–Justices, May 7, 1954, quoted id. at 97.
92. Green v. County School Board, Mr. Justice Brennan, May ___, 1968. Circulated 5-16-68. JMHP.
93. 347 U.S. at 494, note 11.
94. See Schwartz, op. cit. supra note 4, at 704–705.
95. Id. at 705.
96. Ibid.
97. 391 U.S. at 435.
98. Id. at 437–438.
99. Id. at 439.
100. Ibid.
101. Schwartz, op. cit. supra note 4, at 116.
102. Id. at 706.
103. 395 U.S. 225 (1969).
104. Carr v. Montgomery County Board of Education, 232 F. Supp. 705, 707 (M.D. Ala. 1964).
105. Montgomery County Board of Education v. Carr, 400 F.2d 1, 8 (5th Cir. 1968).
106. The Black draft is in JMHP.
107. JMH–HLB, May 21, 1969. JMHP.
108. 395 U.S. at 231.
109. Id. at 235.
110. Compare Wilkinson, op. cit. supra note 29, at 118.
111. Compare id. at 122.
112. Swann v. Charlotte-Mecklenburg Board of Education, 402 U.S. 1, 20 (1971).
113. Graglia, op. cit. supra note 23, at 37.
114. 395 U.S. at 226–227.
115. Id. at 228.
116. Compare ibid.
117. Richmond Times-Dispatch, July 17, 1977, p. A-1.
118. Graglia, op. cit. supra note 23, at 67.
119. Green v. County School Board, 391 U.S. at 439.
120. Compare Wilkinson, op. cit. supra note 29, at 116.
121. Schwartz, op. cit. supra note 4, at 704.
122. 391 U.S. at 437–438.

123. Id. at 439.
124. Id. at 442, n.6.
125. Compare Wilkinson, op. cit. supra note 29, at 117.
126. WJB–WEB, December 30, 1970. JMHP.

Chapter 4

1. 396 U.S. 19 (1969).
2. Brown v. Board of Education, 347 U.S. 483 (1954).
3. Alexander v. Holmes County Board of Education, 90 Sup. Ct. 14 (1969).
4. Ibid.
5. Ibid.
6. See Schwartz, Super Chief: Earl Warren and His Supreme Court—A Judicial Biography 118 (1983).
7. Griffin v. County School Board, 377 U.S. 218, 234 (1964), supra p. 56.
8. Alexander v. Holmes County Board of Education, 90 Sup. Ct. 14, 16 (1969).
9. Id. at 17.
10. Ibid.
11. Alexander v. Holmes County Board of Education, 396 U.S. 802 (1969).
12. See Woodward and Armstrong, The Brethren: Inside the Supreme Court 42 (1979).
13. Brown v. Board of Education, 349 U.S. 294 (1955), supra p. 48.
14. Keyes v. School District No. 1, 90 Sup. Ct. 12 (1969).
15. Unless otherwise stated, the documents quoted in this chapter are in JMHP.
16. 391 U.S. 430 (1968), supra p. 58.
17. Black was referring to Griffin v. County School Board, 377 U.S. 218 (1964), and Green v. County School Board, 391 U.S. 430 (1968).
18. BRW–Chief Justice, October 29, 1969.
19. This copy is not in JMHP.
20. WEB, Memorandum to the Conference, October 27, 1969, writing at end.
21. There were only eight Justices then on the Court, as Justice Fortas, who had resigned, had not yet been replaced.
22. 358 U.S. 1 (1958).
23. Warren, The Memoirs of Earl Warren 298 (1977).
24. This sentence in the Burger draft ended with a comma; but this was obviously a typing error.
25. The Burger draft had the word "the" inserted here. This, too, was an obvious typing error.
26. This copy is not in JMHP.
27. Citing Griffin v. School Board, 377 U.S. 218, 234 (1964), and Green v. County School Board, 391 U.S. 430, 438–439, 442 (1968).
28. BRW–Chief Justice, October 29, 1969.
29. Supra note 22.
30. PS–Chief Justice, October 29, 1969.
31. BRW–Chief Justice, October 29, 1969.
32. November 2, 1969.
33. November 10, 1969, p. 35.
34. For example, St. Louis Post-Dispatch, November 2, 1969, editorial.

Chapter 5

1. 396 U.S. 19 (1969), supra Chapter 4.
2. Green v. County School Board, 391 U.S. 430 (1968).
3. Id. at 439.

4. Alexander v. Holmes County Board of Education, 396 U.S. 1218, 1222 (1969).
5. 396 U.S. at 20 (emphasis added).
6. Supra note 2.
7. United States v. Montgomery County Board of Education, 395 U.S. 225 (1969).
8. Supra p. 62.
9. Supra p. 72.
10. Supra p. 71.
11. 397 U.S. 232 (1970).
12. Id. at 234.
13. This letter is not in JMHP.
14. The final version, 397 U.S. at 235, is slightly different.
15. Unless otherwise stated, all documents quoted in this chapter are in JMHP.
16. The final version, 397 U.S. at 236–237, is slightly different.
17. Swann v. Charlotte-Mecklenburg Board of Education, 431 F.2d 138 (4th Cir. 1970).
18. Swann v. Charlotte-Mecklenburg Board of Education, 399 U.S. 926 (1970).
19. Swann v. Charlotte-Mecklenburg Board of Education, 402 U.S. 1, 11 (1971).
20. New York Times, March 25, 1970, p. 1.
21. WEB, Memorandum to the Conference, September 2, 1970. A copy of the AP story in the Richmond Times-Dispatch, September 1, 1970, is attached. This and the other documents in this section were not gotten from JMHP.
22. New York Times, April 3, 1954, p. 9.
23. New York Times, November 6, 1961, p. 28.
24. Unless otherwise stated, the quotations from the Swann argument are from 70 Briefs and Arguments of the Supreme Court of the United States: Constitutional Law 602–673 (Kurland and Kasper eds. 1975).
25. 39 U.S. L. Wk. 3157 (1970).
26. Ibid.
27. Id. at 3160.
28. Id. at 3159.
29. The Maddox letter was not gotten from JMHP.

Chapter 6

1. Westin, The Anatomy of a Constitutional Case 125 (1958).
2. Brown v. Board of Education, 347 U.S. 483 (1954).
3. Supra p. 92.
4. Brown v. Board of Education, 349 U.S. 294 (1955).
5. United States v. Montgomery County Board of Education, 395 U.S. 225 (1969).
6. Green v. County School Board, 391 U.S. 430 (1968).
7. Supra p. 60.
8. Kemp v. Beasley, 423 F.2d 851, 857 (8th Cir. 1970).
9. Unless otherwise stated, all documents quoted in this chapter are in JMHP.

Chapter 7

1. Quoted in Woodward and Armstrong, The Brethren: Inside the Supreme Court 179–180 (1979).
2. Quoted in Schwartz, Super Chief: Earl Warren and His Supreme Court—A Judicial Biography 522 (1983).
3. Brown v. Board of Education, 347 U.S. 483 (1954). Unless otherwise stated, all documents quoted in this chapter are in JMHP.
4. Green v. County School Board, 391 U.S. 430 (1968).

5. United States v. Montgomery County Board of Education, 395 U.S. 225 (1969).
6. Brown v. Board of Education, 349 U.S. 294 (1955).
7. Alexander v. Holmes County Board of Education, 396 U.S. 19 (1969).
8. Supra p. 51.
9. Supra note 5.
10. Supra note 3.
11. Supra note 4.
12. Learned Hand–FF, November 1, 1958. FFLC.
13. Supra note 3.
14. Supra note 7.
15. Supra note 6.
16. Supra note 4.

Chapter 8

1. Unless otherwise stated, all documents quoted in this chapter are in JMHP.
2. Brown v. Board of Education, 347 U.S. 483 (1954); 349 U.S. 294 (1955).
3. Green v. County School Board, 391 U.S. 430 (1968).
4. United States v. Montgomery County Board of Education, 395 U.S. 225 (1969).
5. McCulloch v. Maryland, 4 Wheat. 316, 407 (U.S. 1819).
6. Supra p. 124.
7. Supra p. 120.

Chapter 9

1. This note is not in JMHP; the Stewart draft dissent is. Unless otherwise stated, all other documents quoted in this chapter are in JMHP.
2. Brown v. Board of Education, 347 U.S. 483 (1954); 349 U.S. 294 (1955).
3. Green v. County School Board, 391 U.S. 430 (1968).
4. United States v. Montgomery County Board of Education, 395 U.S. 225 (1969).
5. Supra p. 51.
6. Ibid.

Chapter 10

1. Green v. County School Board, 391 U.S. 430 (1968).
2. Supra p. 51.
3. Unless otherwise stated, all documents quoted in this chapter are in JMHP.
4. Brown v. Board of Education, 347 U.S. 483 (1954); 349 U.S. 294 (1955).
5. Supra p. 154.
6. 163 U.S. 537 (1896).
7. Id. at 551.
8. This document is not in JMHP.
9. Supra note 1.

Chapter 11

1. TM–WEB, March 23, 1971. JMHP. Unless otherwise stated, all documents quoted in this chapter are in JMHP.
2. Green v. County School Board, 391 U.S. 430 (1968).
3. Brown v. Board of Education, 347 U.S. 483 (1954).
4. This document is not in JMHP.

Chapter 12

1. New York Times, April 6, 1955, p. 22.
2. Id., June 16, 1958, p. 15.
3. 347 U.S. 483 (1954).
4. For a full description, see Schwartz, Super Chief: Earl Warren and His Supreme Court—A Judicial Biography 101 et seq. (1983).
5. Newsweek, May 3, 1971, p. 26.
6. Id. at 27.
7. New Republic, May 1, 1971, p. 12.
8. Swann v. Charlotte-Mecklenburg Board of Education, 402 U.S. 1, 32 (1971).
9. Charlotte Observer, April 21, 1971, p. 1A.
10. Swann v. Charlotte-Mecklenburg Board of Education, 402 U.S. 1, 15 (1971).
11. Charlotte Observer, April 21, 1971, p. 2A.
12. Winston-Salem/Forsyth Board of Education v. Stott, 404 U.S. 1221, 1231, 1228 (1971).
13. Id. at 1228–1229.
14. Swann v. Charlotte-Mecklenburg Board of Education, 402 U.S. 1, 25 (1971) (emphasis added).
15. Loc. cit. supra note 12, at 1221, 1229–1230.
16. See N.Y. Times, September 9, 1971, p. 1.
17. Time, May 3, 1971, p. 40.
18. Supra p. 21.
19. Ibid.
20. N.Y. Times, June 26, 1981, p. A27.
21. Quoted, ibid.
22. Three Cities that are Making Desegregation Work, Report of a National Education Association Special Study 24 (1984); Charlotte Observer, July 12, 1975, p. 18A.
23. NEA, op. cit. supra note 22, at 29.
24. Id. at 34.
25. Id. at 37.
26. Id. at 34.
27. See Swann v. Charlotte-Mecklenburg Board of Education, 67 F.R.D. 648, 649 (W.D.N.C. 1975).
28. Id. at 649, 650.
29. NEA, op. cit. supra note 22.
30. Supra note 3.
31. N.Y. Times, May 17, 1984, p. B18.
32. Id., November 10, 1983, p. D27.
33. Charlotte Observer, April 21, 1971, p. 1A.

APPENDIX A

Chief Justice Burger's December 8, 1970
Memorandum and First Draft *Swann* Opinion

Supreme Court of the United States
Washington, D. C. 20543

CHAMBERS OF
THE CHIEF JUSTICE

December 8, 1970

Re: No. 281—*Swann* v. *Charlotte-Mecklenburg Board of Education*
No. 349—*Charlotte-Mecklenburg Board of Education* v. *Swann*

MEMORANDUM TO THE CONFERENCE:

I enclose typewritten draft of proposed opinion in the above case.

I am sure it is not necessary to emphasize the importance of our attempting to reach an accommodation and a common position, and I would urge that we consult or exchange views by memorandum, or both. Separate opinions, expressing divergent views or conclusions will, I hope, be deferred until we have exhausted all other efforts to reach a common view. I am sure we must all agree that the problems of remedy are at least as difficult and important as the great Constitutional principle of *Brown.*

Regards,

WEB

From: The Chief Justice

Circulated: **DEC 8 1970**

Recirculated: ――――――

No. 281—*Swann v. Charlotte-Mecklenburg Board of Education*
No. 349—*Charlotte-Mecklenburg Board of Education v. Swann*

This case has been in litigation since 1965 and we granted certiorari to review important and unresolved issues as to the duties of the school authorities and the scope of powers of the federal courts under this Court's mandates to eliminate racially separate public schools established and maintained by state action. *Brown v. Board of Education*, 347 U.S. 483 (1954).

This case and those argued with it[1] arose in states having a long history of maintaining two sets of schools in a single school system deliberately operated to carry out a governmental policy to separate pupils in schools solely on the basis of race. That was what *Brown v. Board of Education* was all about. Notwithstanding the frequency of review of cases following *Brown I,* however, the Court has not resolved the specific issues raised in these cases.

I

The Charlotte-Mecklenburg school system encompasses the city of Charlotte and surrounding Mecklenburg County, South Carolina. The area is large—550 square miles—spanning roughly 22 miles east-west and 36 miles north-south. During the 1968–69 school year the system served more than 84,000 pupils in 107 schools. Approximately 71% of the pupils were found to be white and 29% Negro. As of June, 1969, there were approximately 24,000 Negro students in the system, of which 21,000 attended schools within the city of Charlotte. Two-thirds of those 21,000—approximately 14,000 Negro students—attended 21 schools which were either totally Negro or more than 99% Negro.

This situation came about under a desegregation plan approved by the District Court at the commencement of the present litigation in 1965, 243 F.Supp. 667 (WDNC), *affirmed,* 369 F.2d 29 (CA 4, 1966), based upon geographic zoning with a free transfer provision. The present proceedings were initiated in September 1968 by petitioners' motion for

―――――――――

[1] Cite to the companion cases.

further relief based on *Green v. County School Board*, 391 U.S. 430 (1968), and its companion cases.[2] All parties now agree that in 1969 the system fell short of achieving the unitary school system that those cases require.

The District Court held numerous hearings and received voluminous evidence. In addition to finding certain actions of the school board to be discriminatory, the court also found that residential patterns in the city and county resulted in part from federal, state, and local government action other than school board decisions. School board action based on these patterns, for example, by locating schools in Negro residential areas and fixing the size of the schools to accommodate the needs of immediate neighborhoods, resulted in segregated education. These findings were subsequently accepted by the Court of Appeals.

In April, 1969, the District Court ordered the School Board to come forward with a plan for both faculty and student desegregation. Proposed plans were accepted by the court on an interim basis only in June and August 1969. and the board was ordered to file a third plan by November, 1969. In November the Board moved for an extension of time until February, 1970, but when that was denied the Board submitted a partially completed plan. In December, 1969, the District Court held that the Board's submission was unacceptable and appointed an expert in education administration, Dr. John Finger, to prepare a desegregation plan.[3] Thereafter in February, 1970, the District Court was presented with two alternative pupil assignment plans—the finalized "board plan" and the "Finger plan."

The Board Plan. As finally submitted, the School Board plan closed seven schools and reassigned their pupils primarily to increase racial mixing. It severely gerrymandered school attendance zones to promote integration and paired contiguous zones, and non-contiguous zones. It created a single athletic league, eliminated the previously racial basis of the school bus system, integrated faculty and administrative staff, and modified its free transfer plan into a free majority-to-minority transfer system.

The Board plan substantially integrated nine of the system's ten high schools, producing 17% to 36% Negro population in each. The projected Negro attendance at the tenth school, Independence, was 2%.

---

[2] *Raney v. Board of Education*, 391 U.S. 443 (1968), and *Monroe v. Board of Commissioners*, 391 U.S. 450 (1968).

[3] Although Dr. Finger had previously appeared as a witness for one of the parties, the Court of Appeals found that this posture did not cause him to be faithless to the trust the court imposed on him and held the error of his dual role, if any, harmless. We adopt that view but with it the caveat as to the future. Expert witnesses with an interest in sustaining their own plans and positions are placed in an awkward position at best.

The attendance zones for the high schools were typically shaped like wedges of a pie, extending outward from the center of the city to the suburban and rural areas of the county in order to afford residents of the center city area access to outlying schools.

As for junior high schools, the Board Plan rezoned the 21 school areas so that in 20 the Negro attendance would range from 0% to 38%. The other school, located in the heart of the Negro residential area, was left with an enrollment of 90% black.

The board plan with respect to elementary schools relied entirely upon gerrymandering of geographic zones. More than half of the Negro elementary pupils were left in nine schools that were 86% to 100% Negro; approximately half of the white elementary pupils were assigned to schools 86% to 100% white.

The Finger Plan. The plan submitted by the court-appointed expert, Dr. Finger, adopted the school board zoning plan for senior high schools with one modification; it required that an additional 300 Negro students be transported from the Negro residential area of the city to the nearly all-white Independence High School.

The Finger plan for the junior high schools employed much of the rezoning plan of the Board, combined with the creation of nine "satellite" zones.[4] Under the satellite plan, inner-city Negro students were assigned by attendance zones to nine outlying predominantly white junior high schools, thereby substantially integrating every junior high school in the system.

The Finger plan departed from the Board plan chiefly in its handling of the system's 76 elementary schools. Rather than relying solely upon geographic zoning, Dr. Finger proposed use of zoning, pairing, and grouping techniques, with the result that student bodies throughout the system would range from 9% to 38% Negro.[5]

District Court described the plan thus:

> Like the Board plan, the Finger plan does as much by rezoning school attendance lines as can reasonably be accomplished. However, unlike the board plan, it does not stop there. It goes further and desegregates all the rest of the elementary schools by the technique of grouping two or three

[4] A "satellite zone" is an area which is not contiguous with the main attendance zone surrounding the school.

[5] In its opinion and order of December 1, 1969, later incorporated in the order appointing Dr. Finger as consultant, the District Court stated: "Fixed ratios of pupils in particular schools will not be set. If the board in one of its three tries had presented a plan for desegregation, the court would have sought ways to approve variations in pupil ratios. In default of such a plan from the school board, the court will start with the thought . . . *that efforts should be made to reach a 71–29 ratio in the various schools so that there will be no basis for contending that one school is racially different from the others,* but to understand that variations from that norm may be unavoidable." [105a]

outlying schools with one black inner city school; by transporting black students from grades one through four to the outlying white schools; and by transporting white students from the fifth and sixth grades from the outlying white schools to the inner city black school.

Under the Finger plan, nine inner-city Negro schools were grouped in this manner with 24 suburban white schools.

On February 5, 1970, the District Court adopted the Board plan, as modified by Dr. Finger, for the junior and senior high schools. The court rejected the Board elementary school plan and adopted the Finger plan as presented. Implementation was partially stayed by the Court of Appeals for the Fourth Circuit on March 5, and this Court declined to disturb the Fourth Circuit's order, 397 U.S. 978 (1970).

On appeal the Court of Appeals affirmed the District Court's order as to faculty desegregation and the secondary school plans, but vacated the order respecting elementary schools. While agreeing that the District Court properly disapproved the Board plan concerning these schools, the Court of Appeals feared that the pairing and grouping of elementary schools would place an unreasonable burden on the Board and the system's pupils. The case was remanded to the District Court for reconsideration and submission of further plans. This Court granted certiorari, 399 U.S. 926, and directed reinstatement of the District Court's order pending further proceedings in that court.

On remand the District Court received two new plans for elementary schools: a plan prepared by the United States Department of Health, Education and Welfare (The HEW plan) based on contiguous grouping and zoning of schools, and a plan prepared by four members of the nine-member School Board (the minority plan) achieving substantially the same results as the Finger plan but apparently with slightly less transportation. A majority of the School Board declined to amend its proposal. After a lengthy evidentiary hearing the District Court concluded that its own plan (the Finger plan), the minority plan and an earlier draft of the Finger plan were all reasonable and acceptable. It directed the Board to adopt one of the three or in the alternative to come forward with a new, equally effective plan of its own; the court ordered that the Finger plan would remain in effect in the event the School Board declined to adopt a new plan. On August 7, the Board having taken no action, the District Court ordered the Finger plan to remain in effect.

II

It is now 16 years since the Court held, in explicit terms, that state-imposed segregation by race in public schools denies equal protection of the laws. Since then the Court has not deviated and does not now deviate in the slightest degree from that holding or the basic constitutional underpinnings supporting it. Indeed, no hint of challenge to *Brown I* is advanced by any party; those principles are fixed. The implementation of *Brown I* is all that is presented now.

The ramifications and complexities of a shift from two separate and distinct sets of schools within one system was seen by the Court in *Brown II* as one of great difficulty due to myriad variations in hundreds if not thousands of counties and school districts. The record before the Court in *Brown I* had been directed primarily at the basic constitutional right, those cases being class actions affecting thousands of pupils in numerous states. The formulation of decrees to afford appropriate relief was left to subsequent stages. To obtain the maximum assistance of the parties in shaping the remedy the cases were restored to the docket with directions that the parties, including the Attorney General of the United States as *amicus curiae, and* Attorneys General of all states[6] address themselves to the questions relating to remedies.[7]

After arguments and briefs addressed to these questions, the Court, in *Brown II,* reaffirmed its position that all provisions of federal, state or local law requiring discrimination in education based on race violate the Constitution. 349 U.S. 294 (1955). The *Brown II* remand of the cases to the respective district courts having jurisdiction directed attention to specific points. In substance they were:

(1) the primary responsibility of the school authorities to identify and develop remedies for violations existing under varied local school situations;

[6] Six states, Florida, North Carolina, Arkansas, Oklahoma, Maryland and Texas, filed briefs and took part in the oral argument.

[7] Those questions were:

"4 (a) would a decree necessarily follow providing that, within the limits set by normal geographic school districting, Negro children should forthwith be admitted to schools of their choice, or

"(b) may this Court, in the exercise of its equity powers, permit an effective gradual adjustment to be brought about from existing segregated systems to a system not based on color distinctions?

"5 On the assumption on which question 4 (a) and (b) are based, and assuming further that this court will exercise its equity powers to the end described in question 4 (b),

"(a) should this Court formulate detailed decrees in these cases;

"(b) if so, what specific issues should the decrees reach;

"(c) should this Court appoint a special master to hear evidence with a view to recommending specific terms for such decrees;

"(d) should this Court remand to the courts of first instance with directions to frame decrees in these cases, and if so what general directions should the decrees of this Court include and what procedures should the courts of first instance follow in arriving at the specific terms of more detailed decrees?"

347 U.S. 495–496 n. 13.

(2) the continuing obligation of federal courts to determine at the local level whether school authorities were acting in good faith and in accord with controlling constitutional principles;

(3) the shaping of practical flexible remedial decrees so as to reconcile and adjust public and private needs in accord with equitable principles;

(4) the need for non-discriminatory school admission as soon as practicable, taking into account the administrative problems relating to school facilities, transportation systems, and revision of school districts and zones into compact units.

In retrospect it is not surprising that many unanticipated difficulties were encountered. Nothing in our national experience prepared anyone for dealing with changes and adjustment of the magnitude and complexity thereafter to be encountered. Not the least was resistance to the Court's mandates impeding the good faith efforts of some to bring school systems into compliance. The detail and nature of these dilatory aspects are not now relevant. It is sufficient to note that by the time the Court considered *Green* in 1968, very little progress had been made in many areas where dual school systems had historically been maintained by operation of state laws. In *Green,* the Court was confronted with a record of a freedom-of-choice program that the District Court had found to operate in fact to preserve a dual system more than a decade after *Brown II.* While acknowledging that a freedom-of-choice concept could be a valid remedial measure in some circumstances, its failure to be effective in *Green* required that

"The burden on a school board today is to come forward with a plan that promises realistically to work . . . *now* . . . until it is clear that state imposed segregation has been completely removed." *Green,* at 439.

The 1969 Term of Court brought fresh evidence both of the complexities and difficulties and of the dilatory tactics of many school authorities. *Alexander* v. *Holmes County Board of Education,* 396 U.S. 19, again stated the basic obligation emphasized in *Green* that the remedy must be implemented *forthwith.*

In part because of long delay in giving effect to the Court's 1954 and 1955 holdings, the instant cases present for the first time a range of problems encountered by school authorities and the district courts in implementing *Brown I* and *II.* Meanwhile district courts and courts of appeal have struggled in hundreds of cases with a multitude and variety of problems under this Court's general directive. Understandably, those courts had to improvise and experiment without detailed or specific

214 / Appendix A

guidelines. This Court, in *Brown I*, appropriately dealt with the large constitutional principle; other Federal courts had to grapple with the flinty, intractable realities of day-to-day implementation of those constitutional commands. Their efforts, of necessity, embraced a process of "trial and error," and our formulation of guidelines must take into account their experience.

The cases now before us afford the first opportunity to attempt some steps toward guidelines for school authorities and the district courts.[8] The massive basic problem of converting from the state-enforced discrimination inherent in operating separate school systems based on race have been made more complicated since 1954 by changes in the structure and patterns of communities, the growth of student population,[9] movement of families, and other changes, some of which had marked impact on school planning, sometimes neutralizing or negating remedial action before it was implemented.

Moreover, as remedial efforts proceeded differences emerged between the problems of urban and rural schools and between schools of large metropolitan areas and smaller cities. Rural areas accustomed for half a century to the consolidated school systems implemented by bus transportation could make adjustments more readily than metropolitan areas with dense and shifting population, numerous schools, congested and complex traffic patterns. Thus apart from the political, social and economic problems presented by the holding of *Brown I*, school systems have been confronted with numerous practical difficulties no one could foresee in 1954.

The questions now presented reflect, in part at least, the results of sustained efforts by the federal courts to identify the crucial problems and to experiment with a variety of remedies to terminate the operation of two separate school systems within each school district.

Essentially these questions are:

(a) whether the Constitution authorizes courts to require a particular racial balance in each school within a previously racially segregated system;

(b) whether every all-Negro and all-white school must be eliminated as part of a remedial process of desegregation;

---

[8] The necessity for this is suggested by the situation in the 5th Circuit where 166 appeals in school desegregation cases were heard between December 2, 1969, and September 24, 1970.

[9] Elementary public school population (grade 1–6) grew from 17,447,000 in 1954 to 23,103,000 in 1969; secondary school population grew from 11,183,000 in 1954 to 20,775,000 in 1969. Digest of Education Statistics, 1964 ed. p. 6, Office of Education Publication #10024-64. Digest of Education Statistics, 1970 ed. table 28, Office of Education Publication #10024-70.

(c) what are the limits, if any, on the rearrangement of school districts and attendance zones, as a remedial measure; and
(d) what are the limits, if any, on the use of transportation facilities to correct state-enforced racial school segregation.

III

It may be helpful to restate the essential holding of *Brown I* since a proper understanding of that holding is the predicate for any guidelines that can be formulated. It may well be that some of the problems we now face arise from viewing *Brown I* as imposing a requirement for racial balance, *i.e.*, integration, rather than a prohibition against segregation. No holding of this Court has ever required assignment of pupils to establish racial balance or quotas.[10]

The evil struck down by *Brown I*, as contrary to the Equal Protection guarantees of the Constitution, was segregation of public school pupils on a racial basis. This was the violation sought to be corrected by the remedial measures of *Brown II*. We are concerned in these cases with the elimination of the discrimination of the dual school systems, not with the myriad factors of human existence which can cause discrimination in a multitude of ways on racial, religious or ethnic grounds. The evil of discrimination in other areas of life must be dealt with by other remedial mechanisms based on constitutional or statutory guarantees. The heart and core of the cases from *Brown I* to the present embraces two basic elements: (a) separation by race in public schools; (b) enforcement of that separation by governmental action. The elimination of racial discrimination in public schools is a large enough burden and that process will be retarded, not advanced, by efforts to use it for broader purposes not within the power of school authorities. Too much baggage can break down any vehicle.

[10] *Green,* and other opinions seem to have been read over broadly by some as a mandate for integration; fully desegregated schools will, of course, tend to bring about integration. The former is constitutionally required and the latter a probable consequence.
    Similarly, *United States v. Montgomery County Board of Education,* 395 U.S. 225 (1969), is important although it does not relate to desegregation or assignment of pupils, but of faculty. As noted in *Montgomery* there was no claim—and hence, no holding—
    "that racially balanced faculties are constitutionally or legally required." 395 U.S., at 236.
    The *Montgomery* opinion did not go beyond a holding that racial balancing of faculties was a permissible remedial tool to facilitate desegregation. Whatever the legal foundations of the mandate to achieve a "racial balance" in the faculties of each school it is readily apparent on its face that the movement and transportation of teachers by reassignment presents very different problems from those concerning pupils. In metropolitan school systems there is no nexus between the neighborhood or community residence of teachers and the schools to which they are assigned. Basically, of course, racially segregated faculties were part and parcel of the dual school system.

Our objective now and in these cases is to see that school authorities exclude no pupil of a racial minority from any school—directly or indirectly[11]—on account of race; but it does not and cannot embrace all the residential problems, employment patterns, location of public housing, or other factors beyond the jurisdiction of school authorities that may indeed contribute to some disproportionate racial concentration in some schools.

IV

Once a right and a violation have been shown, the scope of equitable remedies to redress past wrongs is broad for in the nature of equitable remedies breadth and flexibility are essential, as *Brown II* emphasized. However as with any equity case, the nature of the school authority's violation determines the scope of the remedy. Courts have little difficulty when confronted with a classical equity claim. An example is found in the hornbook illustration of a farmer who erects a dam in the bed of a stream and cuts off the water supply for neighbors down stream. Upon a proper showing the remedy is simple: a court of equity will command the offending dam to be removed. Similarly an equity court, upon finding an illegal corporate acquisition, can order a divestiture within a given period.

Here, however, we are not confronted with a simple classical equity case, and the simplistic, hornbook remedies are not necessarily relevant. Populations, pupils or misplaced schools cannot be moved as simply as earth by a bulldozer, or property by corporations. The offenders in the cases before us are not the pupils but rather their forebears who erected a systematic barrier to keep Negroes from attending schools with others. The ultimate remedy commanded by *Brown II*, restated and reinforced in numerous intervening cases up to *Alexander,* was to discontinue the dual *system*. Discontinuing separate schools for two racial groups would inevitably bring about schools that were not white or Negro "but just schools." *Green,* at 442. The consequence would be a single integrated system functioning on the same basis as school systems in which no discrimination had ever been enforced.

In seeking at this stage to define even in broad and general terms how far this remedial power extends it is important to remember a factor often easily overlooked in the stress of litigation, that judicial powers are not

[11] The proscription extends, for example, to separation resulting from the manner in which zones are established and the place where a school is to be located. Determining the location of a school in a way that significantly favors one race over others would fall within the proscription. Similarly, transfer policies or practices that foster separation violate the right. *Green, supra.*

co-extensive with those of school authorities. Remedial judicial authority does not put judges automatically in the shoes of school authorities whose powers are plenary. An illustration may be found in the hypothesis of a school authority decision that as a matter of sound educational policy schools should be racially balanced on the declared premise that this is desirable or necessary in order to prepare children for the obligations of citizenship in a pluralistic society. This would appear to be an acceptable educational policy for school authorities based on a permissible foundation and within their plenary powers, and although we decide nothing on this, it is difficult to see what federal challenge could be successfully asserted.

Having said this much, however, it is important to make clear that equity remedies are not uniformly or necessarily limited to what could be done by a court if no pattern of enforced discrimination—past or present—were shown. Although federal courts obviously have no roving, at-large powers to deal with educational policy of the state and local governments, the necessity for a remedy of past violations clothes a court of equity with broad powers.

(1) Racial Balances or Racial Quotas

The constant theme and thrust of every holding from *Brown I* to date is that state-enforced separation of races in public schools violates the Equal Protection Clause. The remedy commanded was to discontinue dual school systems. In this case it is urged that the District Court has imposed a racial balance requirement of 71%–29% on individual schools. The fact that no such objective was actually achieved, and would appear to be impossible, tends to blunt that claim, yet in the opinion and order of the District Court of December 1, 1969, we find the court directing: ". . . that efforts should be made to reach a 71-29 ratio in the various schools so that there will be no basis for contending that one school is racially different from the others. . . ."

The District Judge went on to acknowledge that variation *"from that norm* may be unavoidable." (Emphasis supplied.) This contains strong intimations that the "norm" is a fixed mathematical racial balance reflecting the pupil constituency of the system and this is underscored by the explanation that this ratio is to be achieved "so that there will be no basis for contending that one school is racially different from the others."

Neither the Constitution nor equitable principles grants to judges the power to command that each school in a system reflects, either precisely or substantially, the racial origins of the pupils within the system. Even cursory examination of the records of these cases reveal how transitory a yardstick the school census is at any given time.[12] The fact that school systems maintain a pupil census and often must do so with racial identification, affords no basis for a fixed racial balance; obviously that criterion would require frequent reassignments to reflect population changes.[13]

A district judge confronted with claims of discriminatory segregation of pupils would consider, among other evidentiary matters, the racial composition of the system and the school assignments complained of. Hence, a finding that the total pupil population is, for example, 71% of one race and 29% of another, is one relevant step in identifying a possible violation of the rights of a class of litigants; but the process does not stop there. A court must then determine whether the school authorities assigned these pupils on the capricious and impermissible basis of racial origin to perpetuate segregation. It is not a primary function of school authorities, in the abstract, to construct a system with a racial balance to offset, for example, the imbalances resulting from the residential patterns of the area served even assuming they could do so as a matter of educational policy. That is not compelled by the Constitution.

By the same token it is not the function of a court, lacking as it does the plenary policy powers of a school authority, to order the individual schools to reflect the composition of the system. That result would indeed render the school authorities impervious to a claim ". . . that one school is racially different from the others. . . ." but there is no requirement that this be done.

(2) Must All One-Race Schools be Eliminated?

The records in these cases, as in so many, reveal that in metropolitan areas minority groups are often found to be concentrated in one part of the city.

Despite the most valiant efforts of school authorities and courts, in some circumstances some schools will remain all or largely of one race until new schools can be provided or neighborhood patterns change. Provisions for optional transfer of those in the majority racial group to other schools where they will be in the minority will tend to relieve particular hardship cases for those who find the posture of being part of a

12 For example, in its Memorandum of October 5, 1970, the District Court noted the population of certain school attendance zones had changed because of the movement of families and, in particular, that the racial balance in one school was about to be drastically upset by the opening of a single new low-cost housing project.

13 In its order of February 5, 1970, the District Court instructed the school authorities to "adopt and implement a *continuing* program, computerized or otherwise, of assigning pupils and teachers *during* the school year *as well as at the start of each year* for the conscious purpose of maintaining each school and each faculty in a condition of desegregation." (Emphasis added.) The court's supplemental order of October 5, 1970, gave notice to the school authorities that a new reassignment of pupils was again necessary.

school majority a "badge of inferiority." District courts may provide for optional transfers. Undesirable though it may be, we find nothing in the Constitution, read in its broadest implications, that precludes the maintenance of schools, all or predominantly all of one racial composition in a city of mixed population, so long as the school assignment is not part of state-enforced school segregation.

(3) Remedial Altering of Attendance Zones

The maps submitted in these cases graphically demonstrate that one of the principal tools employed by school planners and courts to break up the dual school system has been a frank—and indeed drastic—gerrymandering of school districts and attendance zones. An additional step was pairing "clustering" or "grouping" of schools with attendance assignments made deliberately to accomplish the transfer of Negro students out of formerly segregated Negro schools and transfer of white students to formerly all Negro schools. More often than not, these zones are not compact or contiguous; indeed they may be on opposite ends of the city. As an interim corrective measure, this cannot be said to be beyond the broad remedial powers of a court in all circumstances.

Absent a history of a dual school system there would be no basis for judicially ordering assignment of students on a racial basis. All things being equal, with no such history of discrimination, it might well be desirable to assign pupils to schools nearby their homes. But all things are not equal in a system that has been deliberately constructed to enforce racial segregation. The remedy for such segregation may be administratively awkward, inconvenient and even bizarre in some situations and may impose burdens on some; but all awkwardness and inconvenience cannot be avoided in the interim period when remedial adjustments are being made solely for the purpose of eliminating the dual school system.

No fixed or even substantially fixed guidelines can be established as to how far a court can go, but it must be recognized that there are limits. The scope and nature of the violation may itself serve some function in defining the remedy.

In this area, we must of necessity rely to a large extent, as this Court has for 16 years, on the informed judgment of the district courts, guided to the extent guidance can be given by our holdings. Perhaps any guidelines are inescapably negative, i.e., that no fixed racial balance, school by school, is required and that gerrymandering and arrangement of attendance zones and transportation programs have limits.

The pairing and grouping of non-contiguous school zones is not an impermissible tool but every judicial step in shaping such zones that goes beyond combinations of contiguous areas should be closely examined. In this respect maps do not tell the whole story; two or even three school zones, on their face widely apart, may be more accessible in terms of travel time because of traffic patterns and good highways than schools geographically closer together.

(4) Transportation of Students

The scope of permissible transportation of students as an implement of a remedial decree has never been defined by this Court. Here again no rigid guidelines can be given for application to the infinite variety of problems presented in thousands of situations.

The objective should be to achieve as nearly as possible that distribution of students and those patterns of assignments that would have normally existed had the school authorities not previously practiced discrimination. This is not to be read as blanket approval of all "racially neutral" assignment plans proposed by school authorities to a district court, for such plans may fail to counteract the continuing effects of past segregation resulting from discriminatory location of school sites or distortion of school size in order to achieve or maintain an artificial racial separation.[14] When school authorities present a district court with a "loaded game board," affirmative action in the form of student transportation may properly be used to achieve truly non-discriminatory assignments. In short, an assignment plan is not acceptable simply because it appears to be neutral.

One primary limitation on the use of transportation as a remedial measure can therefore be found in the nature and extent of past violations by school authorities. But this is not the only limitation for the courts must also weigh the burdens upon students and the potential frustration of legitimate educational goals. It hardly needs expression that the limits on time of travel will vary with many factors, but with none more than the age of the students. The reconciliation of the competing values of the need to grant a remedy for violation of constitutional rights, on one hand, and the well-being of children innocent of wrongdoing, on the other, is a difficult task. It calls for the wisdom of Solomon and the patience of Job; and by and large district judges have exhibited these qualities in the painful period of transition.

The search for solutions is not aided by simplistic slogans for or against "bussing," as though the term described a uniformly invidious course of action. Bus transportation has been used to carry pupils to schools since

---

[14] An artificial racial separation is to be distinguished from separation flowing normally from residential patterns.

the expanded use of the consolidated school systems a half century ago; it will very likely always be so used. An objection to bus transportation of students may have validity, however, when the time or distance of travel is so great as to risk either the health of the children or impinge on the educational process.[15] District courts must weigh the soundness of any transportation plan in light of what is said in subdivisions (1), (2), and (3) above, each of which is a limitation on the use of bus transportation of students.

The Court of Appeals, searching for a term to define the limits on the equitable remedial power of the District Court, employed the concept of "reasonableness" as a test. In *Green, supra,* this Court used the term "feasible" and at least by implication, "workable" and "realistic," in the mandate to develop "a plan that promises realistically to work, and . . . to work *now.*" Words to define the scope of remedial power or the limits on remedial power of courts in an area as sensitive as we deal with here, are poor instruments to convey the sense of basic fairness inherent in equity. We have sought to suggest the nature of limitations without frustrating the appropriate scope of equity.

The cases are remanded for reconsideration by the District Court and further action not inconsistent with this opinion.

---

[15] The contention of the school authorities that 42 U.S.C. §§ 2000c(b) and 2000c–6(a)(2) forbid the transporting of students under a district court order is well answered by the appendix to the separate opinion of Mr. Justice Douglas.

APPENDIX B

Final *Swann* Opinion

MR. CHIEF JUSTICE BURGER delivered the opinion of the Court.

We granted certiorari in this case to review important issues as to the duties of school authorities and the scope of powers of federal courts under this Court's mandates to eliminate racially separate public schools established and maintained by state action. *Brown v. Board of Education*, 347 U. S. 483 (1954) (*Brown I*).

This case and those argued with it[1] arose in States having a long history of maintaining two sets of schools in a single school system deliberately operated to carry out a governmental policy to separate pupils in schools solely on the basis of race. That was what *Brown v. Board of Education* was all about. These cases present us with the problem of defining in more precise terms than heretofore the scope of the duty of school authorities and district courts in implementing *Brown I* and the mandate to eliminate dual systems and establish unitary systems at once. Meanwhile district courts and courts of appeals have struggled in hundreds of cases with a multitude and variety of problems under this Court's general directive. Understandably, in an area of evolving remedies, those courts had to improvise and experiment without detailed or specific guidelines. This Court, in *Brown I*, appropriately dealt with the large constitutional principles; other federal courts had to grapple with the flinty, intractable realities of day-to-day implementation of those constitutional commands. Their efforts, of necessity, embraced a process of "trial and error," and our effort to formulate guidelines must take into account their experience.

---

[1] *McDaniel v. Barresi*, No. 420, *post*, p. 39; *Davis v. Board of School Commissioners of Mobile County*, No. 436, *post*, p. 33; *Moore v. Charlotte-Mecklenburg Board of Education*, No. 444, *post*, p. 47; *North Carolina State Board of Education v. Swann*, No. 498, *post*, p. 43. For purposes of this opinion the cross-petitions in Nos. 281 and 349 are treated as a single case and will be referred to as "this case."

I

The Charlotte-Mecklenburg school system, the 43d largest in the Nation, encompasses the city of Charlotte and surrounding Mecklenburg County, North Carolina. The area is large—550 square miles—spanning roughly 22 miles east-west and 36 miles north-south. During the 1968–1969 school year the system served more than 84,000 pupils in 107 schools. Approximately 71% of the pupils were found to be white and 29% Negro. As of June 1969 there were approximately 24,000 Negro students in the system, of whom 21,000 attended schools within the city of Charlotte. Two-thirds of those 21,000—approximately 14,000 Negro students—attended 21 schools which were either totally Negro or more than 99% Negro.

This situation came about under a desegregation plan approved by the District Court at the commencement of the present litigation in 1965, 243 F. Supp. 667 (WDNC), aff'd, 369 F. 2d 29 (CA4 1966), based upon geographic zoning with a free-transfer provision. The present proceedings were initiated in September 1968 by petitioner Swann's motion for further relief based on *Green* v. *County School Board*, 391 U. S. 430 (1968), and its companion cases.[2] All parties now agree that in 1969 the system fell short of achieving the unitary school system that those cases require.

The District Court held numerous hearings and received voluminous evidence. In addition to finding certain actions of the school board to be discriminatory, the court also found that residential patterns in the city and county resulted in part from federal, state, and local government action other than school board decisions. School board action based on these patterns, for example, by locating schools in Negro residential areas and fixing the size of the schools to accommodate the needs of immediate neighborhoods, resulted in segregated education. These findings were subsequently accepted by the Court of Appeals.

In April 1969 the District Court ordered the school board to come forward with a plan for both faculty and student desegregation. Proposed plans were accepted by the court in June and August 1969 on an interim basis only, and the board was ordered to file a third plan by November 1969. In November the board moved for an extension of time until February 1970, but when that was denied the board submitted a partially completed plan. In December 1969 the District Court held that the board's submission was unacceptable and appointed an

[2] *Raney* v. *Board of Education,* 391 U. S. 443 (1968), and *Monroe* v. *Board of Commissioners,* 391 U. S. 450 (1968).

expert in education administration, Dr. John Finger, to prepare a desegregation plan. Thereafter in February 1970, the District Court was presented with two alternative pupil assignment plans—the finalized "board plan" and the "Finger plan."

The Board Plan. As finally submitted, the school board plan closed seven schools and reassigned their pupils. It restructured school attendance zones to achieve greater racial balance but maintained existing grade structures and rejected techniques such as pairing and clustering as part of a desegregation effort. The plan created a single athletic league, eliminated the previously racial basis of the school bus system, provided racially mixed faculties and administrative staffs, and modified its free-transfer plan into an optional majority-to-minority transfer system.

The board plan proposed substantial assignment of Negroes to nine of the system's 10 high schools, producing 17% to 36% Negro population in each. The projected Negro attendance at the 10th school, Independence, was 2%. The proposed attendance zones for the high schools were typically shaped like wedges of a pie, extending outward from the center of the city to the suburban and rural areas of the county in order to afford residents of the center city area access to outlying schools.

As for junior high schools, the board plan rezoned the 21 school areas so that in 20 the Negro attendance would range from 0% to 38%. The other school, located in the heart of the Negro residential area, was left with an enrollment of 90% Negro.

The board plan with respect to elementary schools relied entirely upon gerrymandering of geographic zones. More than half of the Negro elementary pupils were left in nine schools that were 86% to 100% Negro; approximately half of the white elementary pupils were assigned to schools 86% to 100% white.

The Finger Plan. The plan submitted by the court-appointed expert, Dr. Finger, adopted the school board zoning plan for senior high schools with one modification: it required that an additional 300 Negro students be transported from the Negro residential area of the city to the nearly all-white Independence High School.

The Finger plan for the junior high schools employed much of the rezoning plan of the board, combined with the creation of nine "satellite" zones.[3] Under the satellite plan, inner-city Negro students were assigned by attendance zones to nine outlying predominately white junior high schools, thereby substantially desegregating every junior high school in the system.

The Finger plan departed from the board plan chiefly in its handling

[3] A "satellite zone" is an area which is not contiguous with the main attendance zone surrounding the school.

of the system's 76 elementary schools. Rather than relying solely upon geographic zoning, Dr. Finger proposed use of zoning, pairing, and grouping techniques, with the result that student bodies throughout the system would range from 9% to 38% Negro.[4]

The District Court described the plan thus:

> "Like the board plan, the Finger plan does as much by rezoning school attendance lines as can reasonably be accomplished. However, unlike the board plan, it does not stop there. It goes further and resegregates all the rest of the elementary schools by the technique of grouping two or three outlying schools with one black inner city school; by transporting black students from grades one through four to the outlying white schools; and by transporting white students from the fifth and sixth grades from the outlying white schools to the inner city black school."

Under the Finger plan, nine inner-city Negro schools were grouped in this manner with 24 suburban white schools.

On February 5, 1970, the District Court adopted the board plan, as modified by Dr. Finger, for the junior and senior high schools. The court rejected the board elementary school plan and adopted the Finger plan as presented. Implementation was partially stayed by the Court of Appeals for the Fourth Circuit on March 5, and this Court declined to disturb the Fourth Circuit's order, 397 U. S. 978 (1970).

On appeal the Court of Appeals affirmed the District Court's order as to faculty desegregation and the secondary school plans, but vacated the order respecting elementary schools. While agreeing that the District Court properly disapproved the board plan concerning these schools, the Court of Appeals feared that the pairing and grouping of elementary schools would place an unreasonable burden on the board and the system's pupils. The case was remanded to the District Court for reconsideration and submission of further plans. 431 F. 2d 138. This Court granted certiorari, 399 U. S. 926, and directed reinstatement of the District Court's order pending further proceedings in that court.

On remand the District Court received two new plans for the elementary schools: a plan prepared by the United States Department of Health, Education, and Welfare (the HEW plan) based on contiguous

[4] In its opinion and order of December 1, 1969, later incorporated in the order appointing Dr. Finger as consultant, the District Court stated:

"Fixed ratios of pupils in particular schools will not be set. If the board in one of its three tries had presented a plan for desegregation, the court would have sought ways to approve variations in pupil ratios. In default of any such plan from the school board, the court will start with the thought . . . that efforts should be made to reach a 71–29 ratio in the various schools so that there will be no basis for contending that one school is racially different from the others, but to understand that variations from that norm may be unavoidable." 306 F. Supp. 1299, 1312.

grouping and zoning of schools, and a plan prepared by four members of the nine-member school board (the minority plan) achieving substantially the same results as the Finger plan but apparently with slightly less transportation. A majority of the school board declined to amend its proposal. After a lengthy evidentiary hearing the District Court concluded that its own plan (the Finger plan), the minority plan, and an earlier draft of the Finger plan were all reasonable and acceptable. It directed the board to adopt one of the three or in the alternative to come forward with a new, equally effective plan of its own; the court ordered that the Finger plan would remain in effect in the event the school board declined to adopt a new plan. On August 7, the board indicated it would "acquiesce" in the Finger plan, reiterating its view that the plan was unreasonable. The District Court, by order dated August 7, 1970, directed that the Finger plan remain in effect.

II

Nearly 17 years ago this Court held, in explicit terms, that state-imposed segregation by race in public schools denies equal protection of the laws. At no time has the Court deviated in the slightest degree from that holding or its constitutional underpinnings. None of the parties before us challenges the Court's decision of May 17, 1954, that

> "in the field of public education the doctrine of 'separate but equal' has no place. Separate educational facilities are inherently unequal. Therefore, we hold that the plaintiffs and others similarly situated . . . are, by reason of the segregation complained of, deprived of the equal protection of the laws guaranteed by the Fourteenth Amendment. . . .
>
> "Because these are class actions, because of the wide applicability of this decision, and because of the great variety of local conditions, the formulation of decrees in these cases presents problems of considerable complexity." *Brown* v. *Board of Education, supra*, at 495.

None of the parties before us questions the Court's 1955 holding in *Brown II,* that

> "School authorities have the primary responsibility for elucidating, assessing, and solving these problems; courts will have to consider whether the action of school authorities constitutes good faith implementation of the governing constitutional principles. Because of their proximity to local conditions and the possible need for further hearings, the courts which originally heard these cases can best perform this judicial appraisal. Accordingly, we believe it appropriate to remand the cases to those courts.
>
> "In fashioning and effectuating the decrees, the courts will be guided

by equitable principles. Traditionally, equity has been characterized by a practical flexibility in shaping its remedies and by a facility for adjusting and reconciling public and private needs. These cases call for the exercise of these traditional attributes of equity power. At stake is the personal interest of the plaintiffs in admission to public schools as soon as practicable on a nondiscriminatory basis. To effectuate this interest may call for elimination of a variety of obstacles in making the transition to school systems operated in accordance with the constitutional principles set forth in our May 17, 1954, decision. Courts of equity may properly take into account the public interest in the elimination of such obstacles in a systematic and effective manner. But it should go without saying that the vitality of these constitutional principles cannot be allowed to yield simply because of disagreement with them." *Brown* v. *Board of Education,* 349 U. S. 294, 299–300 (1955).

Over the 16 years since *Brown II,* many difficulties were encountered in implementation of the basic constitutional requirement that the State not discriminate between public school children on the basis of their race. Nothing in our national experience prior to 1955 prepared anyone for dealing with changes and adjustments of the magnitude and complexity encountered since then. Deliberate resistance of some to the Court's mandates has impeded the good-faith efforts of others to bring school systems into compliance. The detail and nature of these dilatory tactics have been noted frequently by this Court and other courts.

By the time the Court considered *Green* v. *County School Board,* 391 U. S. 430, in 1968, very little progress had been made in many areas where dual school systems had historically been maintained by operation of state laws. In *Green,* the Court was confronted with a record of a freedom-of-choice program that the District Court had found to operate in fact to preserve a dual system more than a decade after *Brown II.* While acknowledging that a freedom-of-choice concept could be a valid remedial measure in some circumstances, its failure to be effective in *Green* required that:

"The burden on a school board today is to come forward with a plan that promises realistically to work . . . *now* . . . until it is clear that state-imposed segregation has been completely removed." *Green, supra,* at 439.

This was plain language, yet the 1969 Term of Court brought fresh evidence of the dilatory tactics of many school authorities. *Alexander* v. *Holmes County Board of Education,* 396 U. S. 19, restated the basic obligation asserted in *Griffin* v. *School Board,* 377 U. S. 218, 234 (1964), and *Green, supra,* that the remedy must be implemented *forthwith.*

The problems encountered by the district courts and courts of appeals make plain that we should now try to amplify guidelines, however incomplete and imperfect, for the assistance of school authorities and courts.[5] The failure of local authorities to meet their constitutional obligations aggravated the massive problem of converting from the state-enforced discrimination of racially separate school systems. This process has been rendered more difficult by changes since 1954 in the structure and patterns of communities, the growth of student population,[6] movement of families, and other changes, some of which had marked impact on school planning, sometimes neutralizing or negating remedial action before it was fully implemented. Rural areas accustomed for half a century to the consolidated school systems implemented by bus transportation could make adjustments more readily than metropolitan areas with dense and shifting population, numerous schools, congested and complex traffic patterns.

III

The objective today remains to eliminate from the public schools all vestiges of state-imposed segregation. Segregation was the evil struck down by *Brown I* as contrary to the equal protection guarantees of the Constitution. That was the violation sought to be corrected by the remedial measures of *Brown II*. That was the basis for the holding in *Green* that school authorities are "clearly charged with the affirmative duty to take whatever steps might be necessary to convert to a unitary system in which racial discrimination would be eliminated root and branch." 391 U. S., at 437–438.

If school authorities fail in their affirmative obligations under these holdings, judicial authority may be invoked. Once a right and a violation have been shown, the scope of a district court's equitable powers to remedy past wrongs is broad, for breadth and flexibility are inherent in equitable remedies.

"The essence of equity jurisdiction has been the power of the Chancellor to do equity and to mould each decree to the necessities of the particular case. Flexibility rather than rigidity has distinguished it. The qualities of mercy and practicality have made equity the instrument for nice adjust-

[5] The necessity for this is suggested by the situation in the Fifth Circuit where 166 appeals in school desegregation cases were heard between December 2, 1969, and September 24, 1970.

[6] Elementary public school population (grades 1–6) grew from 17,447,000 in 1954 to 23,103,000 in 1969; secondary school population (beyond grade 6) grew from 11,183,000 in 1954 to 20,775,000 in 1969. Digest of Educational Statistics, Table 3, Office of Education Pub. 10024–64; Digest of Educational Statistics, Table 28, Office of Education Pub. 10024–70.

ment and reconciliation between the public interest and private needs as well as between competing private claims." *Hecht Co.* v. *Bowles*, 321 U. S. 321, 329–330 (1944), cited in *Brown II, supra*, at 300.

This allocation of responsibility once made, the Court attempted from time to time to provide some guidelines for the exercise of the district judge's discretion and for the reviewing function of the courts of appeals. However, a school desegregation case does not differ fundamentally from other cases involving the framing of equitable remedies to repair the denial of a constitutional right. The task is to correct, by a balancing of the individual and collective interests, the condition that offends the Constitution.

In seeking to define even in broad and general terms how far this remedial power extends it is important to remember that judicial powers may be exercised only on the basis of a constitutional violation. Remedial judicial authority does not put judges automatically in the shoes of school authorities whose powers are plenary. Judicial authority enters only when local authority defaults.

School authorities are traditionally charged with broad power to formulate and implement educational policy and might well conclude, for example, that in order to prepare students to live in a pluralistic society each school should have a prescribed ratio of Negro to white students reflecting the proportion for the district as a whole. To do this as an educational policy is within the broad discretionary powers of school authorities; absent a finding of a constitutional violation, however, that would not be within the authority of a federal court. As with any equity case, the nature of the violation determines the scope of the remedy. In default by the school authorities of their obligation to proffer acceptable remedies, a district court has broad power to fashion a remedy that will assure a unitary school system.

The school authorities argue that the equity powers of federal district courts have been limited by Title IV of the Civil Rights Act of 1964, 42 U. S. C. § 2000c. The language and the history of Title IV show that it was enacted not to limit but to define the role of the Federal Government in the implementation of the *Brown I* decision. It authorizes the Commissioner of Education to provide technical assistance to local boards in the preparation of desegregation plans, to arrange "training institutes" for school personnel involved in desegregation efforts, and to make grants directly to schools to ease the transition to unitary systems. It also authorizes the Attorney General, in specified circumstances, to initiate federal desegregation suits. Section 2000c (b) defines "desegregation" as it is used in Title IV:

" 'Desegregation' means the assignment of students to public schools and within such schools without regard to their race, color, religion, or national origin, but 'desegregation' shall not mean the assignment of students to public schools in order to overcome racial imbalance."

Section 2000c–6, authorizing the Attorney General to institute federal suits, contains the following proviso:

"nothing herein shall empower any official or court of the United States to issue any order seeking to achieve a racial balance in any school by requiring the transportation of pupils or students from one school to another or one school district to another in order to achieve such racial balance, or otherwise enlarge the existing power of the court to insure compliance with constitutional standards."

On their face, the sections quoted purport only to insure that the provisions of Title IV of the Civil Rights Act of 1964 will not be read as granting new powers. The proviso in § 2000c–6 is in terms designed to foreclose any interpretation of the Act as expanding the *existing* powers of federal courts to enforce the Equal Protection Clause. There is no suggestion of an intention to restrict those powers or withdraw from courts their historic equitable remedial powers. The legislative history of Title IV indicates that Congress was concerned that the Act might be read as creating a right of action under the Fourteenth Amendment in the situation of so-called "de facto segregation," where racial imbalance exists in the schools but with no showing that this was brought about by discriminatory action of state authorities. In short, there is nothing in the Act that provides us material assistance in answering the question of remedy for state-imposed segregation in violation of *Brown I*. The basis of our decision must be the prohibition of the Fourteenth Amendment that no State shall "deny to any person within its jurisdiction the equal protection of the laws."

IV

We turn now to the problem of defining with more particularity the responsibilities of school authorities in desegregating a state-enforced dual school system in light of the Equal Protection Clause. Although the several related cases before us are primarily concerned with problems of student assignment, it may be helpful to begin with a brief discussion of other aspects of the process.

In *Green*, we pointed out that existing policy and practice with regard to faculty, staff, transportation, extracurricular activities, and facilities were among the most important indicia of a segregated system. 391 U. S. at 435. Independent of student assignment, where it is possible to identify a "white school" or a "Negro school" simply by reference to the racial composition of teachers and staff, the quality of school buildings and equipment, or the organization of sports activities, a *prima facie* case of violation of substantive constitutional rights under the Equal Protection Clause is shown.

When a system has been dual in these respects, the first remedial responsibility of school authorities is to eliminate invidious racial distinctions. With respect to such matters as transportation, supporting personnel, and extracurricular activities, no more than this may be necessary. Similar corrective action must be taken with regard to the maintenance of buildings and the distribution of equipment. In these areas, normal administrative practice should produce schools of like quality, facilities, and staffs. Something more must be said, however, as to faculty assignment and new school construction.

In the companion *Davis* case, *post*, p. 33, the Mobile school board has argued that the Constitution requires that teachers be assigned on a "color blind" basis. It also argues that the Constitution prohibits district courts from using their equity power to order assignment of teachers to achieve a particular degree of faculty desegregation. We reject that contention.

In *United States* v. *Montgomery County Board of Education*, 395 U. S. 225 (1969), the District Court set as a goal a plan of faculty assignment in each school with a ratio of white to Negro faculty members substantially the same throughout the system. This order was predicated on the District Court finding that:

> "The evidence does not reflect any real administrative problems involved in immediately desegregating the substitute teachers, the student teachers, the night school faculties, and in the evolvement of a really legally adequate program for the substantial desegregation of the faculties of all schools in the system commencing with the school year 1968–69." Quoted at 395 U. S., at 232.

The District Court in *Montgomery* then proceeded to set an initial ratio for the whole system of at least two Negro teachers out of each 12 in any given school. The Court of Appeals modified the order by eliminating what it regarded as "fixed mathematical" ratios of faculty and substituted an initial requirement of *"substantially* or *approximately"* a five-to-one ratio. With respect to the future, the Court of Appeals held that the numerical ratio should be eliminated and that compliance should not be tested solely by the achievement of specified proportions. *Id.*, at 234.

We reversed the Court of Appeals and restored the District Court's order in its entirety, holding that the order of the District Judge "was adopted in the spirit of this Court's opinion in *Green* . . . in that his plan 'promises realistically to work, and promises realistically to work *now*.' The modifications ordered by the panel of the Court of Appeals, while of course not intended to do so, would, we think, take from the order some of its capacity to expedite, by means of specific commands, the day when a completely unified, unitary, nondiscriminatory school system becomes a reality instead of a hope. . . . We also believe that under all the circumstances of this case we follow the original plan outlined in *Brown II* . . . by accepting the more specific and expeditious order of [District] Judge Johnson . . . ." 395 U. S., at 235–236 (emphasis in original).

The principles of *Montgomery* have been properly followed by the District Court and the Court of Appeals in this case.

The construction of new schools and the closing of old ones are two of the most important functions of local school authorities and also two of the most complex. They must decide questions of location and capacity in light of population growth, finances, land values, site availability, through an almost endless list of factors to be considered. The result of this will be a decision which, when combined with one technique or another of student assignment, will determine the racial composition of the student body in each school in the system. Over the long run, the consequences of the choices will be far reaching. People gravitate toward school facilities, just as schools are located in response to the needs of people. The location of schools may thus influence the patterns of residential development of a metropolitan area and have important impact on composition of inner-city neighborhoods.

In the past, choices in this respect have been used as a potent weapon for creating or maintaining a state-segregated school system. In addition to the classic pattern of building schools specifically intended for Negro or white students, school authorities have sometimes, since *Brown*, closed schools which appeared likely to become racially mixed through changes in neighborhood residential patterns. This was sometimes accompanied by building new schools in the areas of white suburban expansion farthest from Negro population centers in order to maintain the separation of the races with a minimum departure from the formal principles of "neighborhood zoning." Such a policy does more than simply influence the short-run composition of the student body of a new school. It may well promote segregated residential patterns which, when combined with "neighborhood zoning," further lock the school system into the mold of separation of the races. Upon a proper showing a district court may consider this in fashioning a remedy.

In ascertaining the existence of legally imposed school segregation, the existence of a pattren of school construction and abandonment is thus a factor of great weight. In devising remedies where legally imposed segregation has been established, it is the responsibility of local authorities and district courts to see to it that future school construction and abandonment are not used and do not serve to perpetuate or re-establish the dual system. When necessary, district courts should retain jurisdiction to assure that these responsibilities are carried out. Cf. *United States* v. *Board of Public Instruction*, 395 F. 2d 66 (CA5 1968); *Brewer* v. *School Board*, 397 F. 2d 37 (CA4 1968).

V

The central issue in this case is that of student assignment, and there are essentially four problem areas:

(1) to what extent racial balance or racial quotas may be used as an implement in a remedial order to correct a previously segregated system;
(2) whether every all-Negro and all-white school must be eliminated as an indispensable part of a remedial process of desegregation;
(3) what the limits are, if any, on the rearrangement of school districts and attendance zones, as a remedial measure; and
(4) what the limits are, if any, on the use of transportation facilities to correct state-enforced racial school segregation.

(1) Racial Balances or Racial Quotas

The constant theme and thrust of every holding from *Brown I* to date is that state-enforced separation of races in public schools is discrimination that violates the Equal Protection Clause. The remedy commanded was to dismantle dual school systems.

We are concerned in these cases with the elimination of the discrimination inherent in the dual school systems, not with myriad factors of human existence which can cause discrimination in a multitude of ways on racial, religious, or ethnic grounds. The target of the cases from *Brown I* to the present was the dual school system. The elimination of racial discrimination in public schools is a large task and one that should not be retarded by efforts to achieve broader purposes lying beyond the jurisdiction of school authorities. One vehicle can carry only a limited amount of baggage. It would not serve the important objective of *Brown I* to seek to use school desegregation cases for purposes beyond their scope, although desegregation of schools ultimately will have impact on other forms of discrimination. We do not reach in this case the question whether a showing that school segregation is a consequence of other types of state action, without any discriminatory action by the school authorities, is a constitutional violation requiring remedial action by a school desegregation decree. This case does not present that question and we therefore do not decide it.

Our objective in dealing with the issues presented by these cases is to see that school authorities exclude no pupil of a racial minority from any school, directly or indirectly, on account of race; it does not and cannot embrace all the problems of racial prejudice, even when those problems contribute to disproportionate racial concentrations in some schools.

In this case it is urged that the District Court has imposed a racial balance requirement of 71%–29% on individual schools. The fact that no such objective was actually achieved—and would appear to be impossible—tends to blunt that claim, yet in the opinion and order of the District Court of December 1, 1969, we find that court directing

> "that effort should be made to reach a 71–29 ratio in the various schools so that there will be no basis for contending that one school is racially different from the others . . . , [t]hat no school [should] be operated with an all-black or predominantly black student body, [and] [t]hat pupils of all grades [should] be assigned in such a way that as nearly as practicable the various schools at various grade levels have about the same proportion of black and white students."

The District Judge went on to acknowledge that variation "from that norm may be unavoidable." This contains intimations that the "norm" is a fixed mathematical racial balance reflecting the pupil constituency of the system. If we were to read the holding of the District Court to require, as a matter of substantive constitutional right, any particular degree of racial balance or mixing, that approach would be disapproved and we would be obliged to reverse. The constitutional command to desegregate schools does not mean that every school in every community must always reflect the racial composition of the school system as a whole.

As the voluminous record in this case shows,[7] the predicate for the District Court's use of the 71%–29% ratio was twofold: first, its express finding, approved by the Court of Appeals and not challenged here, that a dual school system had been maintained by the school authorities at least until 1969; second, its finding, also approved by the Court of Appeals, that the school board had totally defaulted in its acknowledged

[7] It must be remembered that the District Court entered nearly a score of orders and numerous sets of findings, and for the most part each was accompanied by a memorandum opinion. Considering the pressure under which the court was obliged to operate we would not expect that all inconsistencies and apparent inconsistencies could be avoided. Our review, of course, is on the orders of February 5, 1970, as amended, and August 7, 1970.

duty to come forward with an acceptable plan of its own, notwithstanding the patient efforts of the District Judge who, on at least three occasions, urged the board to submit plans.[8] As the statement of facts shows, these findings are abundantly supported by the record. It was because of this total failure of the school board that the District Court was obliged to turn to other qualified sources, and Dr. Finger was designated to assist the District Court to do what the board should have done.

We see therefore that the use made of mathematical ratios was no more than a starting point in the process of shaping a remedy, rather than an inflexible requirement. From that starting point the District Court proceeded to frame a decree that was within its discretionary powers, as an equitable remedy for the particular circumstances.[9] As we said in *Green*, a school authority's remedial plan or a district court's remedial decree is to be judged by its effectiveness. Awareness of the racial composition of the whole school system is likely to be a useful starting point in shaping a remedy to correct past constitutional violations. In sum, the very limited use made of mathematical ratios was within the equitable remedial discretion of the District Court.

(2) One-race Schools

The record in this case reveals the familiar phenomenon that in metropolitan areas minority groups are often found concentrated in one part of the city. In some circumstances certain schools may remain all or largely of one race until new schools can be provided or neighborhood patterns change. Schools all or predominately of one race in a district of mixed population will require close scrutiny to determine that school assignments are not part of state-enforced segregation.

In light of the above, it should be clear that the existence of some small number of one-race, or virtually one-race, schools within a district

8 The final board plan left 10 schools 86% to 100% Negro and yet categorically rejected the techniques of pairing and clustering as part of the desegregation effort. As discussed below, the Charlotte board was under an obligation to exercise every reasonable effort to remedy the violation, once it was identified, and the suggested techniques are permissible remedial devices. Additionally, as noted by the District Court and Court of Appeals, the board plan did not assign white students to any school unless the student population of that school was at least 60% white. This was an arbitrary limitation negating reasonable remedial steps.

9 In its August 3, 1970, memorandum holding that the District Court plan was "reasonable" under the standard laid down by the Fourth Circuit on appeal, the District Court explained the approach taken as follows:

"This court has not ruled, and does not rule that 'racial balance' is required under the Constitution; nor that all black schools in all cities are unlawful; nor that all school boards must bus children or violate the Constitution; nor that the particular order entered in this case would be correct in other circumstances not before this court." (Emphasis in original.)

is not in and of itself the mark of a system that still practices segregation by law. The district judge or school authorities should make every effort to achieve the greatest possible degree of actual desegregation and will thus necessarily be concerned with the elimination of one-race schools. No *per se* rule can adequately embrace all the difficulties of reconciling the competing interests involved; but in a system with a history of segregation the need for remedial criteria of sufficient specificity to assure a school authority's compliance with its constitutional duty warrants a presumption against schools that are substantially disproportionate in their racial composition. Where the school authority's proposed plan for conversion from a dual to a unitary system contemplates the continued existence of some schools that are all or predominately of one race, they have the burden of showing that such school assignments are genuinely nondiscriminatory. The court should scrutinize such schools, and the burden upon the school authorities will be to satisfy the court that their racial composition is not the result of present or past discriminatory action on their part.

An optional majority-to-minority transfer provision has long been recognized as a useful part of every desegregation plan. Provision for optional transfer of those in the majority racial group of a particular school to other schools where they will be in the minority is an indispensable remedy for those students willing to transfer to other schools in order to lessen the impact on them of the state-imposed stigma of segregation. In order to be effective, such a transfer arrangement must grant the transferring student free transportation and space must be made available in the school to which he desires to move. Cf. *Ellis* v. *Board of Public Instruction,* 423 F. 2d 203, 206 (CA5 1970). The court orders in this and the companion *Davis* case now provide such an option.

(3) *Remedial Altering of Attendance Zones*

The maps submitted in these cases graphically demonstrate that one of the principal tools employed by school planners and by courts to break up the dual school system has been a frank—and sometimes drastic—gerrymandering of school districts and attendance zones. An additional step was pairing, "clustering," or "grouping" of schools with attendance assignments made deliberately to accomplish the transfer of Negro students out of formerly segregated Negro schools and transfer of white students to formerly all-Negro schools. More often than not, these zones are neither compact[10] nor contiguous; indeed they may be on opposite

[10] The reliance of school authorities on the reference to the "revision of . . . attendance areas into *compact* units," *Brown II,* at 300 (emphasis supplied), is misplaced.

ends of the city. As an interim corrective measure, this cannot be said to be beyond the broad remedial powers of a court.

Absent a constitutional violation there would be no basis for judicially ordering assignment of students on a racial basis. All things being equal, with no history of discrimination, it might well be desirable to assign pupils to schools nearest their homes. But all things are not equal in a system that has been deliberately constructed and maintained to enforce racial segregation. The remedy for such segregation may be administratively awkward, inconvenient, and even bizarre in some situations and may impose burdens on some; but all awkwardness and inconvenience cannot be avoided in the interim period when remedial adjustments are being made to eliminate the dual school systems.

No fixed or even substantially fixed guidelines can be established as to how far a court can go, but it must be recognized that there are limits. The objective is to dismantle the dual school system. "Racially neutral" assignment plans proposed by school authorities to a district court may be inadequate; such plans may fail to counteract the continuing effects of past school segregation resulting from discriminatory location of school sites or distortion of school size in order to achieve or maintain an artificial racial separation. When school authorities present a district court with a "loaded game board," affirmative action in the form of remedial altering of attendance zones is proper to achieve truly nondiscriminatory assignments. In short, an assignment plan is not acceptable simply because it appears to be neutral.

In this area, we must of necessity rely to a large extent, as this Court has for more than 16 years, on the informed judgment of the district courts in the first instance and on courts of appeals.

We hold that the pairing and grouping of noncontiguous school zones is a permissible tool and such action is to be considered in light of the objectives sought. Judicial steps in shaping such zones going beyond combinations of contiguous areas should be examined in light of what is said in subdivisions (1), (2), and (3) of this opinion concerning the objectives to be sought. Maps do not tell the whole story since noncontigu-

---

The enumeration in that opinion of considerations to be taken into account by district courts was patently intended to be suggestive rather than exhaustive. The decision in *Brown II* to remand the cases decided in *Brown I* to local courts for the framing of specific decrees was premised on a recognition that this Court could not at that time foresee the particular means which would be required to implement the constitutional principles announced. We said in *Green, supra*, at 439:

"The obligation of the district courts, as it always has been, is to assess the effectiveness of a proposed plan in achieving desegregation. There is no universal answer to complex problems of desegregation; there is obviously no one plan that will do the job in every case. The matter must be assessed in light of the circumstances present and the options available in each instance."

ous schoool zones may be more accessible to each other in terms of the critical travel time, because of traffic patterns and good highways, than schools geographically closer together. Conditions in different localities will vary so widely that no rigid rules can be laid down to govern all situations.

(4) Transportation of Students

The scope of permissible transportation of students as an implement of a remedial decree has never been defined by this Court and by the very nature of the problem it cannot be defined with precision. No rigid guidelines as to student transportation can be given for application to the infinite variety of problems presented in thousands of situations. Bus transportation has been an integral part of the public education system for years, and was perhaps the single most important factor in the transition from the one-room schoolhouse to the consolidated school. Eighteen million of the Nation's public school children, approximately 39%, were transported to their schools by bus in 1969–1970 in all parts of the country.

The importance of bus transportation as a normal and accepted tool of educational policy is readily discernible in this and the companion case, *Davis, supra*.[11] The Charlotte school authorities did not purport to assign students on the basis of geographically drawn zones until 1965 and then they allowed almost unlimited transfer privileges. The District Court's conclusion that assignment of children to the school nearest their home serving their grade would not produce an effective dismantling of the dual system is supported by the record.

Thus the remedial techniques used in the District Court's order were within that court's power to provide equitable relief; implementation of the decree is well within the capacity of the school authority.

The decree provided that the buses used to implement the plan would operate on direct routes. Students would be picked up at schools near their homes and transported to the schools they were to attend. The trips for elementary school pupils average about seven miles and the District Court found that they would take "not over 35 minutes at the

[11] During 1967–1968, for example, the Mobile board used 207 buses to transport 22,094 students daily for an average round trip of 31 miles. During 1966–1967, 7,116 students in the metropolitan area were bused daily. In Charlotte-Mecklenburg, the system as a whole, without regard to desegregation plans, planned to bus approximately 23,000 students this year, for an average daily round trip of 15 miles. More elementary school children than high school children were to be bused, and four- and five-year-olds travel the longest routes in the system.

most."[12] This system compares favorably with the transportation plan previously operated in Charlotte under which each day 23,600 students on all grade levels were transported an average of 15 miles one way for an average trip requiring over an hour. In these circumstances, we find no basis for holding that the local school authorities may not be required to employ bus transportation as one tool of school desegregation. Desegregation plans cannot be limited to the walk-in school.

An objection to transportation of students may have validity when the time or distance of travel is so great as to either risk the health of the children or significantly impinge on the educational process. District courts must weigh the soundness of any transportation plan in light of what is said in subdivisions (1), (2), and (3) above. It hardly needs stating that the limits on time of travel will vary with many factors, but probably with none more than the age of the students. The reconciliation of competing values in a desegregation case is, of course, a difficult task with many sensitive facets but fundamentally no more so than remedial measures courts of equity have traditionally employed.

VI

The Court of Appeals, searching for a term to define the equitable remedial power of the district courts, used the term "reasonableness." In *Green, supra,* this Court used the term "feasible" and by implication, "workable," "effective," and "realistic" in the mandate to develop "a plan that promises realistically to work, and . . . to work *now.*" On the facts of this case, we are unable to conclude that the order of the District Court is not reasonable, feasible and workable. However, in seeking to define the scope of remedial power or the limits on remedial power of courts in an area as sensitive as we deal with here, words are poor instruments to convey the sense of basic fairness inherent in equity. Substance, not semantics, must govern, and we have sought to suggest the nature of limitations without frustrating the appropriate scope of equity.

At some point, these school authorities and others like them should have achieved full compliance with this Court's decision in *Brown I.* The systems would then be "unitary" in the sense required by our decisions in *Green* and *Alexander.*

12 The District Court found that the school system would have to employ 138 more buses than it had previously operated. But 105 of those buses were already available and the others could easily be obtained. Additionally, it should be noted that North Carolina requires provision of transportation for all students who are assigned to schools more than one and one-half miles from their homes. N. C. Gen. Stat. § 115–186 (b) (1966).

It does not follow that the communities served by such systems will remain demographically stable, for in a growing, mobile society, few will do so. Neither school authorities nor district courts are constitutionally required to make year-by-year adjustments of the racial composition of student bodies once the affirmative duty to desegregate has been accomplished and racial discrimination through official action is eliminated from the system. This does not mean that federal courts are without power to deal with future problems; but in the absence of a showing that either the school authorities or some other agency of the State has deliberately attempted to fix or alter demographic patterns to affect the racial composition of the schools, further intervention by a district court should not be necessary.

For the reasons herein set forth, the judgment of the Court of Appeals is affirmed as to those parts in which it affirmed the judgment of the District Court. The order of the District Court, dated August 7, 1970, is also affirmed.

It is so ordered.

Index